Lecture Notes in Control and Information Sciences 205

Editor: M. Thoma

W0107654

Springer-Verlag London Ltd.

Ülle Kotta

Inversion Method in the Discrete-time Nonlinear Control Systems Synthesis Problems

Springer

Author

Ülle Kotta, DSc
Institute of Cybernetics, Akadeemia tee 21, Tallinn, Estonia

ISBN 978-3-540-19966-3 Springer-Verlag Berlin Heidelberg New York

British Library Cataloguing in Publication Data
A catalogue record for this book is available from the British Library

Typesetting: Camera ready by author

69/3830-543210 Printed on acid-free paper

Preface

This book is devoted to the study of discrete-time nonlinear control systems whose behaviour is governed by state equations. Our attention is focused on synthesis of the closed-loop systems with prespecified (completely or partially) input-output maps and the systematic approach basing on system inversion has been developed for this purpose.

During the past two decades or so, there has been a great deal of interest in the study of nonlinear control systems as a separable discipline itself. The differential geometric theory of nonlinear control systems is an especially well-developed area, reflected in books by A.Isidori [Nonlinear Control Systems, Springer-Verlag, Berlin, Heidelberg 1985 and 1989] and H.Nijmeijer and A. van der Schaft [Nonlinear Dynamical Control Systems, Springer-Verlag, New York 1990], containing fundamentals of theory of nonlinear control systems.

Our purpose in writing this book is twofold. On the one hand, the aim of this book is to cover nonlinear control systems theory from another viewpoint. Namely, our purpose is to survey the control system design methods, based on system inversion techniques. We call these methods inversion methods. Of course, the scope of our book is much more narrow than those of the above mentioned two books. On the other hand, we want to collect in one place many of the recent results on discrete-time nonlinear systems. Continuous emphasis on the study of discrete-time and sampled-data control systems has been greatly motivated by new generations of low cost and small size digital computers. Though the book by Nijmeijer and van der Schaft contains a chapter devoted to discrete-time control systems, there is not at present a book on the topic of discrete-time nonlinear control systems.

The basic ideas used in this book are the ideas of many people. It is known for a long time that inverse systems either implicitly or explicitly may be used to solve numerous control problems. The inversion problem is of direct interest in control problems such as servo, output tracking and feedforward control. Despite the apparent importance and great conceptual simplicity of the notion of right invertibility there seems to be no monograph written on this topic.

Control system design via the inversion method is a systematic design technique which bases on fundamental property – invertibility property – of the system. There are really two important types of invertibility. The property of left invertibility, in broad terms, means the possibility to compute the input uniquely from the knowledge of the output. The property of right invertibility means the ability of the control system to reproduce an arbitrary function at the output by manipulating the control input. The inversion method is based on the notion of right invertibility and its application reduces to the construction of the right inverse system for the system to be controlled.

For this reason at first we shall consider the problem of right invertibility of the control system in its various aspects: necessary and sufficient conditions for right invertibility, the algorithms for constructing the right inverses, the properties of right inverses etc. Then the possibility to apply the right inverse for designing the control systems with given input-output maps will be considered and the inversion method will be proposed. The author then applies the inversion method to several control problems of long-standing interest: input-output decoupling, input-output linearization, disturbance decoupling, and various conbinations of these problems.

The book consists of two parts. We have tried to present the material and the main ideas in a systematic manner starting from elementary topics in the first part of the book and progressing to areas of current research in the second part of the book. The first part of this book deals with special subset of right invertible systems, the so-called $(d_1, \ldots, d_p)$-forward time-shift right invertible systems. The methods described in Part I are elaborated on in Part II. In the second part of the book the notion of the forward time-shift right invertibility will be introduced in such a way that the notion of $(d_1, \ldots, d_p)$-forward time-shift right invertibility is a special case of the first notion. On the bases of the generalized notion of right invertibility both inversion method and the earlier results on control system design will be extended to the wider class of systems – to forward time-shift right invertible systems, or even only to partly right invertible systems. The book has also an introductory chapter – Chapter 1 – which presents several examples of discrete-time nonlinear control systems, taken from different scientific disciplines and discusses briefly the specific features of discrete-time nonlinear control systems. At the end of the book the open problems will be discussed. In the text the references are not given to the original sources. At the end of each chapter we have added bibliographical notes about main sources we have used, as well as some partial historical information. Furthermore, we have occasionally added some references to related work and further developments.

Though the aim of this book is to survey the inversion method and the results that can be obtained by this method in connection with discrete-time nonlinear control systems, the basic ideas are also applicable to several classes of systems such as continuous-time nonlinear systems, implicit systems, 2-D systems, distributed parameter systems and, of course, to linear systems.

Acknowledgements

Many people own my gratitude without whose assistance and/or motivation this book would not have been written. At first, I wish to acknowledge Prof. M.Fliess whose inspiring papers revealed to me, many years ago, the wonderful roads of theory of nonlinear control systems. I started my research on the topic of discrete-time nonlinear control systems being guided his papers, as well as those of S.Monaco and D.Normand-Cyrot. My special gratitude belongs to Henk Nijmeijer who actually opened me (living at that time in the USSR) the door to the western world. Over the years, cooperation with him and with his colleagues from Twente University has learnt me much. I would like to thank him also for his helpful criticism and suggestions for improvement the manuscript of this book. More recently, I owe much to the contacts with J.Descusse, C.Moog and their colleagues from the Laboratoire d'Automatique de Nantes. Over the years the Institute of Cybernetics (Estonian Academy of Sciences) has offered me excellent surrounding for research. I am much indebted to my former supervisor Prof. Ü.Jaaksoo who aroused my interest in inverse systems. I wish to express my sincerest gratitude to Prof. J.Engelbrecht for his constant encouragement and to Pilvi Veeber for her skilful typing of the manuscript.

Partial support towards the development of the book was provided by Estonian Science Foundation.

The book represents the revised lecture notes which were prepared for a course on discrete-time nonlinear control systems, given for the first time at the Centre of Research and Advanced Studies of the National Polytechnic Institute (CINVESTAV) in Mexico City for M.S. and Ph.D. students in engineering. The author is pleased to be able to express her gratitude to prof. R.Castro and his colleagues from Automatic Control Group for this opportunity.

Table of Contents

1. Introduction

This book is concerned with discrete-time nonlinear control systems described by difference equations of the form

$$\begin{aligned} x(t+1) &= f(x(t), u(t), w(t)), \\ y(t) &= h(x(t), u(t)), \end{aligned} \tag{1.1}$$

where x denotes the state of the system, u and w are the inputs, y is the output. By u is denoted the control input, i.e. the input which we can manipulate and by w is denoted the disturbance input which we cannot manipulate. In some cases the disturbance is unknown, the other cases it can be measured or it is missing altogether. In this book by outputs we mean the variables-to-be-controlled, i.e. the variables in whose behaviour we are especially interested, and not the measurements that we can perform on the system. We assume throughout the book that the state vector x of the system can be measured.

Before we will discuss the main assumptions on the system (1.1) which will be supposed to hold throughout the book and the types of feedback that will be applied to the system (1.1), we focus on several examples of control systems which fit into (1.1). The examples serve as a motivation for considering discrete-time nonlinear control systems. As one will see, the examples are taken from different scientific disciplines such as economics, chemical engineering, biology and medicine.

1.1 Examples of Discrete-time Nonlinear Control Systems

1.1.1 An Example of Discrete-time Nonlinear Control Model in Economics

Discrete-time nonlinear dynamic models arise naturally in economics. Here we present a simple macroeconomic model that describes an economy composed of two sectors:

$$\begin{aligned} x_1(t+1) &= \frac{1}{1+\gamma}[x_1(t) + w(t)f(x_1(t), x_2(t))] - \frac{w(t)}{1+\gamma} f(x_1(t), x_2(t))u_1(t), \\ x_2(t+1) &= \frac{x_2(t)}{1+\gamma} + \frac{w(t)}{1+\gamma} f(x_1(t), x_2(t))u_1(t) + \frac{1-w(t)}{1+\gamma} f(x_1(t), x_2(t))u_2(t). \end{aligned}$$

In the above model by x_1 and x_2 are denoted the capital stocks in the private and the government sectors, respectively; u_1 is the saving tax rate and u_2 is the tax rate of consumption. The rate of saving by the private sector is denoted by $w(t)$ and γ is a constant.

1.1.2 Stagewise Processes in Chemical Engineering

Chemical engineers make extensive use of stagewise processes whose typical examples are plate absorption and distillation columns and a series of batch reaction, extraction or leaching tanks. In each case the operation is characterized by a finite between-stage change in the value of the dependent variable. The finite transition between stages can be best modelled by difference equations. We shall present here a simple example of countercurrent liquid-liquid extraction.

The process of countercurrent liquid-liquid extraction consists of M separate stages. Each stage consists a separate tank. Under steady—state operation, L moles of extract solvent and R moles of raffinate solvent are charged to tank k, mixed, allowed to settle, and the raffinate sent to tank $k-1$ and the extract to tank $k+1$. This process takes place in each tank. A sketch of the system is shown in Fig. 1.1. The initial raffinate charge to tank 1 on each cycle consists of R moles of solvent with a composition $y(0)$ (moles of transferable material A per mole of solvent). The initial extract charge to tank M on each cycle consists of L moles of extract solvent with a composition $z(M+1)$ (moles of transferable material A per mole of solvent). Raffinate and solvent are immiscible. Equilibrium is obtained in each tank. The equilibrium relation is

$$z(k) = \beta y(k) \, . \tag{1.2}$$

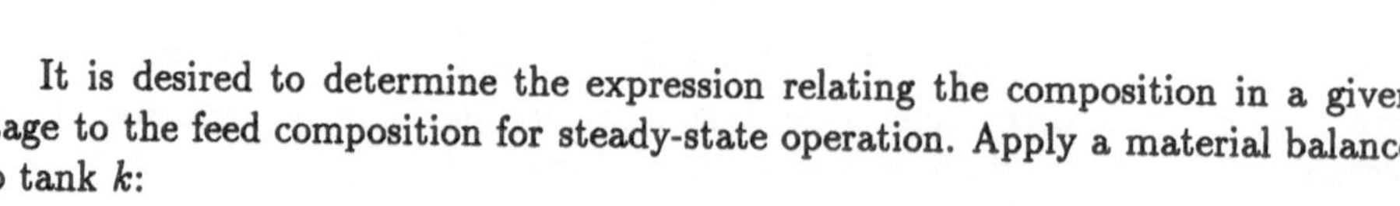

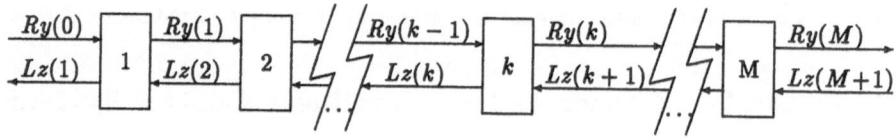

Fig. 1.1. Countercurrent liquid-liquid extraction stages.

It is desired to determine the expression relating the composition in a given stage to the feed composition for steady-state operation. Apply a material balance to tank k:

$$Ry(k-1) + Lz(k+1) = Ry(k) + Lz(k) \tag{1.3}$$

When (1.3) is combined with (1.2) it results

$$y(k+1) = (\alpha + 1)y(k) - \alpha y(k-1), \; k = 0, 1, \ldots, M \, , \tag{1.4}$$

where

$$\alpha = R/(L\beta)$$

can be considered as the control variable u. Denoting

$$y(k) = x_2(k)$$

the difference equation (1.4) may be written in the state space form

$$\begin{aligned} x_1(k+1) &= x_2(k) \\ x_2(k+1) &= (u+1)x_2(k) - ux_1(k)\,, \quad k = 0,1,\ldots,M \\ y(k) &= x_2(k)\,. \end{aligned} \tag{1.5}$$

1.1.3 An Example of Discrete-time Nonlinear Control Model in Medicine

Nonlinear control systems often appear in pharmacokinetics and chemotherapy. There is a certain temptation to employ discrete-time models since in clinical practice and in laboratory, systems are generally observed at discrete times.

The next model describes the behaviour of hyperthyroid patients undergoing antithyroid drug therapy:

$$\begin{aligned} x_1(t+1) &= a_1 \exp[-a_2 x_3(t)] + a_3 x_1(t) \\ x_2(t+1) &= a_4 \exp[-a_5 u(t)] + a_6 x_2(t) \\ x_3(t+1) &= a_7 x_2(t+1) + a_8 x_4(t+1) + a_9 x_3(t) \\ x_4(t+1) &= a_{10} x_2(t+1) + a_{11} x_4(t+1) \\ x_5(t+1) &= a_{12} f(t) + a_{14} x_5(t) \end{aligned}$$

where

$$f(t) = \begin{cases} 0, \text{ if } x_3(t) \le a_{13} \\ 1 - \exp[a_{13} - x_3(t)]\,, \text{ if } x_3(t) > a_{13}\,. \end{cases}$$

In this model the following notations have been used:

$x_1(t)$ – plasma TSH at time t, mU/liter
$x_2(t)$ – stored iodoprotein at time t, nM
$x_3(t)$ – plasma T_3 concentration at time t
$x_4(t)$ – plasma T_4 concentration at time t
$x_5(t)$ – clinical index at time t
u(t) – drug dosage per day at time t, mg/day.

The second, third and fourth equations describe the dynamics of thyroid hormones and in the second equation, drug action on hormone synthesis is included.

1.1.4 An Example of Discrete-time Nonlinear Control Model in Renewable Resource Management

Bioeconomic problems have been recognized as a rich source of mathematical control problems. Ecosystem managing includes maximizing sustainable yield, controlling pests and diseases. There are many situations where a difference equation model is more appropriate than differential equation. This happens when population growth occurs at discrete times and when generations are completely nonoverlapping. Biological examples are provided by some insect populations and certain fishes which live in isolated generations. Discrete-time formulation is also widely used for describing age-structured population models since one ordinarily keeps track of age groups instead of exact age. Very often there is a natural unit of time associated with a particular resource system. In most cases, the structure of the model is a crude approximation to reality and it is difficult to find instances where control may be applied precisely to actual situation. It is common to see the control policies as indicative.

It has been recognized that ecosystem models must include a degree of uncertainty in their description. In real ecosystems, this uncertainty arises due to fluctuations in the environment, or neglecting less important interactions between populations in the ecosystem. Usually, these uncertainties are described via external perturbation vector in the model over which we have no control and the model is exactly (1.1) with $w(t)$ as perturbation vector. The elements x_i of x, $i = 1, \dots, n$ usually denote the density (biomass, number) of the ith species, although in ecosystem models, some of the variables x_i may represent the concentration of nutrients and chemical pollutants. By $u(t)$ is usually denoted the harvesting rate and by $w(t)$ the perturbations which describe uncertainties of the model. One-dimensional state equations describe the growth of one species. The multidimensional models describe the complex interactions between animal and plant populations in the development of the ecosystem. Animal and plant populations compete and cooperate with one another to obtain sufficient natural resources to sustain their growth and survival.

Next model describes the effects of two species on one another. The two species, krill and baleen whale, are related by the fact that one is a source of food for other:

$$\begin{aligned} x_1(t+1) &= x_1(t)\exp\{r_1[1-(x_1(t)+\alpha x_2(t))/(1+\alpha)]\} + b_1u_1(t)x_1(t))\,, \\ x_2(t+1) &= x_2(t)\exp\{r_2[1-(x_2(t)/x_1(t)]\} + b_2u_2(t)x_2(t))\,. \end{aligned}$$

Here $x_1(t)$ and $x_2(t)$ represent the krill (prey) and baleen whale (predator) population, respectively, at time t; $u(t)$ is the harvesting rate at time t, r_1, r_2, b_1, b_2 and α are some constants.

1.1.5 Sampled-data Systems

Most commonly discrete-time nonlinear control systems appear as the discretization of continuous-time nonlinear systems. In the control of continuous-time systems present-day technology often asks for digitally implemented controllers, and sometimes it is preferable to design a controller in the discrete-time domain rather than discretizing the continuous-time control law.

Here an exact step-invariant sampled-data model of a multi-input-multi-output continuous-time system S, linear is control, is given. The considered system is described by differential equations of the form

$$\dot{x} = f_0(x) + \sum_{i=1}^{m} f_i(x)u_i\,, \quad x(0) = x_0\,, \tag{1.6}$$

where $x \in X \subset R^n$, $u = (u_1 \dots u_m) \in R^m$, $f_i : X \to X$, $i = 0, 1, \dots, m$ are analytic functions.

The exact step-invariant sampled-data model of a continuous-time system (1.6) is defined as the one whose response to a step input (i.e. the type of control usually available under digital control)

$$u(kT + t) = u(kT)\,,\; 0 \le t < T$$

is identical to that of the continuous-time system at discrete instants of time kT, $k \ge 0$, provided the initial states of both systems are the same.

Derivation of sampled-data model is based on the representation of the solution of the differential equation (1.6) in terms of a formal Lie exponential series

$$x(kT + t) = \sum_{r \ge 0} \frac{t^r}{r!} L^r_{f_0 + \sum_{i=1}^m f_i u_i} x|_{x(kT)}\,,\; 0 \le t < T \tag{1.7}$$

where by $L_{f_0 + \sum_{i=1}^m f_i u_i}$ is denoted the Lie differential operator associated with the function $f_0(x) + \sum_{i=1}^m f_i(x) u_i$

$$L_{f_0 + \sum_{i=1}^m f_i u_i} = \sum_{j=1}^n \left[f_{0,j}(x) + \sum_{i=1}^m f_{i,j}(x) u_i \right] \frac{\partial}{\partial x_j}\,,$$

by $L^r_{f_0 + \sum_{i=1}^m f_i u_i}$ its r-multiple composition

$$L^r_{f_0 + \sum_{i=1}^m f_i u_i} = L_{f_0 + \sum_{i=1}^m f_i u_i} \left(L^{r-1}_{f_0 + \sum_{i=1}^m f_i u_i} \right)\,,\; r \ge 2\,,$$

and by $L^0_{f_0 + \sum_{i=1}^m f_i u_i}$ an identity operator I. Of course, (1.7) may only be defined for T sufficiently small.

As we are interested in the states x only at sampling instants $kT + T$, we obtain from (1.7) for $t = T$

$$x(kT + T) = \sum_{r \ge 0} \frac{T^r}{r!} L^r_{f_0 + \sum_{i=1}^m f_i u_i} x|_{x(kT)} \triangleq F(x(kT), u(kT))\,,$$

where $F(x, u)$ because of

$$L_{f_0 + \sum_{i=1}^m f_i u_i} = L_{f_0} + \sum_{i=1}^m u_i L_{f_i}$$

takes the following form

$$\begin{aligned} F(x,u) = F_0(x) &+ \sum_{i_1=1}^m F_{i_1}(x) u_{i_1} + \sum_{i_1,i_2=1}^m F_{i_1 i_2}(x) u_{i_1} u_{i_2} + \dots \\ &\dots + \sum_{i_1,\dots,i_s=1}^m F_{i_1 \dots i_s}(x) u_{i_1} \dots u_{i_s} + \dots\,, \end{aligned} \tag{1.8}$$

$$F_{i_1 \dots i_p}(x) = \sum_{r=0}^{\infty} \frac{T^{r+p}}{(r+p)!} \sum_{\substack{c_1 + \dots + c_{p+1} = r \\ c_1 \ge 0, \dots, c_{p+1} \ge 0}} L^{c_1}_{f_0} L_{f_{i_1}} L^{c_2}_{f_0} L_{f_{i_2}} \dots L_{f_{i_p}} L^{c_{p+1}}_{f_0} x\,.$$

The exact step-invariant sampled-data model (1.8) is defined in terms of infinite series both with respect to the sampling period T and the control u. So, in the general case, the model (1.8) is usually not computable. In reality, to compute the model, one must confine oneself with finite number of terms in this series. In that way we reach the notion of approximate sampled-date-models. Computing approximate sampled-date model corresponds to the truncation of the infinite series with respect to the sampling period T as well as to the control u, at the fixed orders τ and λ, respectively, which define the orders of approximation of the sampled system

$$F^{\tau,\lambda}(x,u) = F_0^{\tau}(x) + \sum_{i_1=1}^{m} F_{i_1}^{\tau}(x)u_{i_1} + \ldots + \sum_{i_1,\ldots,i_\lambda=1}^{m} F_{i_1\ldots i_\lambda}^{\tau}(x)u_{i_1}\ldots u_{i_\lambda},$$

$$F_{i_1\ldots i_p}^{\tau}(x) = \sum_{r=0}^{\tau-p} \frac{T^{r+p}}{(r+p)!} \sum_{\substack{c_1+\ldots+c_{p+1}=r \\ c_1\geq 0,\ldots,c_{p+1}\geq 0}} L_{f_0}^{c_1} L_{f_{i_1}} L_{f_0}^{c_2} \ldots L_{f_{i_p}} L_{f_0}^{c_{p+1}} x.$$

The response of the (τ,λ)th-order approximate sampled-date model to a step input will agree with that of the continuous-time system at given instants of time $t = kT$, $k \geq 0$, up to an error of order $O(T^{\tau+1}, u^{\lambda+1})$.

The τth-order approximate sampled-date model can be computed recursively on the bases of the $(\tau-1)$th-order model, starting with the first-order model which is equivalent to the classical Eulel discretization scheme

$$F_{i_1\ldots i_p}^{\tau}(x) = L_{f_0} F_{i_1\ldots i_p}^{\tau-1}(x) + L_{f_{i_1}} F_{i_2\ldots i_p}^{\tau-1}(x). \tag{1.9}$$

1.2 Discrete-time Nonlinear Control Systems

Next we will briefly discuss discrete-time nonlinear control systems given by (1.1), i.e. by the equations

$$x(t+1) = f(x(t), u(t), w(t)),$$
$$y(t) = h(x(t), u(t)),$$

where as before x, u, w and y denote reprectively the state, the control, the disturbance and the output. We shall work on a finite time interval, that means we assume (1.1) to hold for $t = 0, 1, 2, \ldots, t_F$, where t_F is some finite time instant, $t_F < \infty$. We assume that the states $x(t)$ belong to an open subset X of R^n, the controls $u(t)$ belong to an open subset U of R^m, the disturbances $w(t)$ belong to an open subset W of R^r and the outputs $y(t)$ belong to an open subset Y of R^p, all for $0 \leq t \leq t_F$. Then (1.1) is a shorthand writing for

$$\begin{cases} x_1(t+1) = f_1(x_1(t), \ldots, x_n(t), u_1(t), \ldots, u_m(t), w_1(t), \ldots, w_r(t)), \\ \ldots \\ x_n(t+1) = f_n(x_1(t), \ldots, x_n(t), u_1(t), \ldots, u_m(t), w_1(t), \ldots, w_r(t)), \end{cases}$$

$$\begin{cases} y_1(t) = h_1(x_1(t), \ldots, x_n(t), u_1(t), \ldots, u_m(t)), \\ \ldots \\ y_p(t) = h_p(x_1(t), \ldots, x_n(t), u_1(t), \ldots, u_m(t)). \end{cases} \tag{1.10}$$

We assume the state transition map $f : X \times U \times W \to X$ to be smooth mapping. In this context smooth means C^∞, i.e. that all of the partial derivatives $\partial^{s+r+k}/\partial x_{i_1} \ldots \partial x_{i_s} \, \partial u_{j_1} \ldots \partial u_{j_r} \, \partial w_{l_1} \ldots \partial w_{l_k}$ exist and are continuous, though many results which will be given in the next chapters hold under weaker conditions. In many circumstances f only needs to be sufficiently many times continuously differentiable with respect to x, u and w. We work with C^∞ functions in order not to have to keep track of the exact number of derivatives required in different situations. Note that if $f : X \times U \to X$ and $h : X \to Y$ are smooth and X, U, Y are open subsets, then the composition $h \circ f : X \times U \to Y$ is also smooth. Sometimes it will be useful to strengthen the smoothness condition and to require that f is (real) analytic. Similarly we assume the output map $h : X \times U \to Y$ to be smooth or analytic. For convenience we consider the system (1.10), initialized at $t = 0$; since the system is time-invariant (i.e. not explicitly depending on time) this may be done without loss of generality.

The system (1.1) has as inputs the sequences $\mathbf{u} = \{u(t);\ 0 \le t \le t_F\}$ of control vectors from $U \subset R^m$ and sequences $\mathbf{w} = \{w(t);\ 0 \le t \le t_F\}$ of disturbance vectors from $W \subset R^r$. We denote the sets of such time sequences by $\mathcal{U}$ and $\mathcal{W}$ respectively. Together with (1.1) we have to specify the classes of admissible controls $\mathcal{U}$ and admissible disturbances $\mathcal{W}$. For discrete-time system (1.1), $\mathcal{U}$ may be any set of time functions $\mathbf{u} : Z_0^{t_F} \to U$ where $Z_0^{t_F}$ stands for the set $\{0, 1, 2, \ldots, t_F\}$. Similarly, a class of admissible disturbances $\mathcal{W}$ may be any set of time functions $\mathbf{w} : Z_0^{t_F} \to W$.

For difference equation

$$x(t+1) = f(x(t), u(t), w(t)),\ x(0) = x_0,$$

as long as f is a well-defined function on $X \times U \times W$, there is no problem regarding existence and uniqueness of its solution $x(t)$, $0 \le t \le t_F$, for any admissible control sequence $\mathbf{u} \in \mathcal{U}$, any admissible disturbance sequence $\mathbf{w} \in \mathcal{W}$ and any arbitrary initial state $x_0 \in R^n$. Such a solution will usually be denoted as $x(t, x_0, \mathbf{u}, \mathbf{w})$ which is a shorthand writing for $x(t, x_0, u(0), \ldots, u(t), w(0), \ldots, w(t))$. Similarly, we let $y(t, x_0, \mathbf{u}, \mathbf{w}) = y(t, x_0, u(0), \ldots, u(t), w(0), \ldots, w(t))$ denote the trajectory of the output for the system (1.1) corresponding to a choice of the control sequence $u(0), \ldots, u(t)$, the disturbance sequence $w(0), \ldots, w(t)$ and the initial state x_0.

In chapters 2, 3, 5, 6 we shall consider systems without disturbances

$$\begin{aligned} x(t+1) &= f(x(t), u(t)),\ x(0) = x_0, \\ y(t) &= h(x(t), u(t)) \end{aligned} \tag{1.11}$$

where x, u, y and h are defined as before and $f : X \times U \to X$.

The system $S = \{U, Y, X, f, h\}$ given by (1.11) generates for each initial state x_0 the input-output (IO) map $\Sigma_{x_0}^S$ on the set $\mathcal{U}$. The IO map

$$\Sigma_{x_0}^S : \mathcal{U} \to \mathcal{Y}$$

assigns to each input sequence $\mathbf{u} \in \mathcal{U}$ the output sequence $\mathbf{y} = \{y(t) \in Y;\ 0 \le t \le$ $\le t_F\} \in \mathcal{Y}$ according to the recursion (1.11) and assuming $x(0) = x_0$.

If $C = \{U^C, Y^C, X^C, f^C, h^C\}$ is another control system of the form (1.11) with $Y^C \subseteq U$, we may define the series connection of S with C, denoted by $S \circ C$, to be the system

$$S \circ C = \{U^C, Y, X \times X^C, f^{S \circ C}, h^{S \circ C}\},$$

where

$$f^{S \circ C}(x, x^C, u^C) = \begin{bmatrix} f(x, h(x^C, u^C)) \\ f^C(x^C, u^C) \end{bmatrix}$$

and

$$h^{S \circ C}(x, x^C, u) = h(x, u).$$

Thus $S \circ C$ is the system which results when the output of C is made the input to S. The initial state of $S \circ C$ is (x_0, x_0^C) where by x_0^C is denoted the initial state of C.

Clearly, the IO map of the composite system $S \circ C$ is the composition of the IO maps for S and C:

$$\Sigma^{S \circ C}_{(x_0, x_0^C)} = \Sigma^S_{x_0} \circ \Sigma^C_{x_0^C}. \tag{1.12}$$

Throughout the book we shall adopt a local viewpoint. In the discrete-time case the local study is useless around an arbitrary initial state since even in one step, the state can move far away from the initial state, regardless of the small control and disturbance values. In order not to loose localness one possibility is to work around an equilibrium point of the system, that is around $(x^0, u^0, w^0) \in X \times U \times W$ such that $f(x^0, u^0, w^0) = x^0$. Since in general the disturbances $w(t)$, $0 \leq t \leq t_F$ need not be necessarily close to w^0, for systems with disturbances working around an equilibrium point is more restrictive if compared to the case without disturbances. Another, a more general possibility, is to work around certain set of time sequences of state, control, disturbance and output $\{\bar{x}(t), \bar{u}(t), \bar{w}(t), \bar{y}(t);\ 0 \leq t \leq t_F\}$ that satisfy the system equations (1.1). This set of time sequences is called the (reference) trajectory of the system. Note that an equilibrium point can also be considered a trajectory, namely it is a trajectory consisting only in one point. Working around a trajectory is useful in situations where there is no natural equilibrium point or an equilibrium point has little relevance for control purposes. In this book mainly the first approach will be followed but for a few problems local solutions will be presented around a given trajectory.

If we assume to work in a neighbourhood of an equilibrium point and on a finite time interval $0 \leq t \leq t_F$, then, using the control sequence $\{u(t);\ 0 \leq t \leq t_F\}$ with each $u(t)$ sufficiently close to u^0 and provided that in the disturbance sequence $\{w(t);\ 0 \leq t \leq t_F\}$ each $w(t)$ is sufficiently close to w^0 and that the initial state x_0 is sufficiently close to x^0 we can assure that the states $x(t)$ are sufficiently close to x^0, and the outputs $y(t)$ are sufficiently close to $y^0 = h(x^0)$, both for $0 \leq t \leq t_F$.

Let us denote by X^0 the open subset of X such that $\| x - x^0 \| < \gamma$ for some $\gamma > 0$. Let us denote by U^0 the set of controls u which are sufficiently close to u^0, that is $\| u - u^0 \| < \delta$ for some $\delta > 0$. Let us denote by $Y^0 (Y_i^0)$ the set of outputs y (scalar output components y_i) which are sufficiently close to $y^0 (y_i^0)$, that is $\| y - y^0 \| <$ $< \varepsilon$ $(\| y_i - y_i^0 \| < \varepsilon)$ for some $\varepsilon > 0$. Denote by $\mathcal{U}^0$ and $\mathcal{Y}^0 (\mathcal{Y}_i^0)$ the sets of control sequences $\{u(t) \in U^0;\ 0 \le t \le t_F\}$ and output sequences $\{y(t) \in Y^0;\ 0 \le t \le t_F\}$ (output component sequences $\{y_i(t) \in Y_i^0;\ 0 \le t \le t_F\}$), respectively.

1.3 Static and Dynamic State Feedback

In the next chapters we will discuss changing the structure of nonlinear control systems via feedback. Since we assume throughout the book that the whole state vector of the system can be measured, we shall only consider state feedback. We will be interested in two types of state feedback for a system (1.1) or (1.11)—static state feedback and dynamic state feedback.

A static state feedback for the nonlinear system (1.11) or (1.1) is definedas a relation

$$u(t) = \varphi(x(t), v(t)) \tag{1.13}$$

where $v \in R^m$ represents a new control vector. Note that since we are working locally around an equilibrium point (x^0, u^0) of the system (1.11), the feedback (1.13) is also defined only locally around a point (x^0, v^0, u^0), such that v^0 satisfies the relation $u^0 = \varphi(x^0, v^0)$.

A feedback (1.13) is called a regular static state feedback if $\partial\varphi(x,v)/\partial v$ is a nonsingular matrix for all x and v in the domain of φ.

The regularity of feedback implies that the closed loop system, i.e. the system composed from the original system (1.1) and the feedback (1.13)

$$\begin{aligned} x(t+1) &= f(x(t), \varphi(x(t), v(t)), w(t)) \\ y(t) &= h(x(t), \varphi(x(t), v(t))) \end{aligned}$$

admits as many independent controls as the original system.

The adjective "static" indicates the fact that the feedback (1.13) decides what the value of u at a time-instant t should be only on the basis of what the value of x and v at that specific time-instant is. In this sense (1.13) could also be called a memoryless feedback. If we also incorporate some kind of memory in the feedback we arrive at what is called a dynamic state feedback.

A dynamic state feedback for the system (1.11) or (1.1) is defined as

$$\begin{aligned} x^C(t+1) &= f^C(x(t), x^C(t), v(t)) \\ u(t) &= \varphi(x(t), x^C(t), v(t)) \end{aligned} \tag{1.14}$$

where $x^C \in X^C \subset R^{n_C}$, $f^C : X \times X^C \times V \to X^C$, $\varphi : X \times X^C \times V \to U$ are smooth mappings and $v \in V \subset R^m$ represents a new control vector. Here x^C can be interpreted as the memory of feedback. Sometimes the system (1.14) is called a compensator.

A feedback (1.14) is called a regular dynamic state feedback if the system (1.1), (1.14)

$$\begin{aligned} x(t+1) &= f(x(t), \varphi(x(t), x^C(t), v(t)), w(t)) \\ x^C(t+1) &= f^C(x(t), x^C(t), v(t)) \\ u(t) &= \varphi(x(t), x^C(t), v(t)) \end{aligned} \tag{1.15}$$

with inputs $v(t)$ and outputs $u(t)$ defines a one-to-one (x, x^C, w)-dependent correspondence between the input variables v and the output variables u.

In the case when measurements of the disturbances are available, we allow the feedback to depend on disturbance w. So, in this case we shall use a static state feedback of the form

$$u(t) = \varphi(x(t), v(t), w(t))$$

and a dynamic state feedback of the form

$$\begin{aligned} x^C(t+1) &= f^C(x(t), x^C(t), v(t), w(t)) \\ u(t) &= \varphi(x(t), x^C(t), v(t), w(t)) \,. \end{aligned}$$

Notes and References

The economic example has been taken from [Aok75a]. The other examples of discrete-time dynamic models in economics, simple as well more advanced can be found in [Aok75a], [Aok75b], [Aok76], [Nij89], [NS90]. A detailed exposition of the process of countercurrent liquid-liquid extraction was given in [MSR57] where more examples of discrete-time nonlinear stagewise processes are discussed. For a further discussion of the drug therapy model we refer to [Swa84]. The ecosystem model is discussed in [Fis87]. For more ecosystem and renewable resourse management models, simple and advanced see [Cla76], [Fis87], [Get87], [KL88], [Lev84], [May74], [May78], [Ost78], [Pal83], [Pie77], [Sch85].

The problem of finding the exact sampled-data model for continuous-time linear-analytic system was first addressed and solved by Monaco and Normand-Cyrot [MNC85] for single-input systems. The sampled-data model (1.8) ha been taken from [Kot89] and the formulas (1.9) for recursive computation of the approximate sampled-data model from [Kot94]. The different formulas for computing sampled-data models can be found in [MNC86a], [MNC90]. While all formulas for computing sampled-data models are similar in the sense that their derivation is based on the representation of the solution of the continuous-time system in terms of a formal Lie exponential series [FLL83], and that all formulas give the same smpled-data models when applied to any concrete system, they have several differences. The formulas given in [MNC86a], [MNC90] have a form which is better suited for studing several properties of the sampled-data models, such as controllability and observability. However, they are more complicated to apply than (1.8) since they contain combinatorics, shuffle products and *ad*-operators. If one wants just to find the sampled-data model of some concrete process, the easier formulas (1.8) will do the job. An exact sampled-data model for multi-input continuous-time system, nonlinear in control is presented in [Kot94]. In [Bar89] the program written with the aid of computer algebra system REDUCE for computing sampled-data model of linear-analytic system has been presented.

We should like to end up with some remarks concerning the form of the state transition map f in (1.11) (or in (1.1)). No significant simplification occurs when f is linear in control, i.e. when f has the structure commonly used in the continuous time. This is due to the fact that in discrete-time, one extensively uses the operation of composition of functions which generally does not preserve the specific structure [MNC86b]. Moreover, by the results of section 1.1.5, the linear analytic dynamics under sampling are no longer linear in control.

[All91] Allen L.J.S. Discrete and continuous models of populations and epidemics. *Journal of Mathematical Systems, Estimation and Control*, 1991, v. 1, 335–369.

[Aok75a] Aoki M. Some examples of dynamic bilinear models in economics. *Lecture Notes in Economics and Mathematical Systems.* Springer Verlag, Berlin, 1975, v. 111, 163–169.

[Aok75b] Aoki M. On a generalization of Tinbergen's condition in theory of policy to dynamic models. *Review of Economic Studies*, 1975, 42, 293–296.

[Aok76] Aoki M. *Optimal Control and System Theory in Dynamic Economic Analysis.* North-Holland, New York, 1976.

[Bar89] Barbot J.P. A computer-aided design for sampling a nonlinear analytic systems. *Lect. Notes Comp. Sci.*, 1989, v. 357, 74–88.

[Cla76] Clark C.W. *Mathematical Bioeconomics. The Optimal Management of Renewable Resources.* John Wiley, New York, 1976.

[Fis87] Fisher M.E. Variability in ecosystem models: a deterministic approach. *Lect. Notes in Biomathematics*, v. 72, 1987, 139–151.

[FLL83] Fliess M., M.Lamnabhi, and F.Lamnabhi-Lagarrigue. An algebraic approach to nonlinear functional expansions. *IEEE Trans. on Circuits and Systems*, 1983, v. 30, 554–570.

[Get87] Getz W.M. Modeling for biological resource management. *Lect. Notes in Biomathematics*, v. 72, 1987, 22–42.

[KL88] Kaitala V. and G.Leitmann. Control of uncertain discrete systems: an application in resource management. *Proc. 27th IEEE Conference on Decision and Control*, Austin, Texas, 1988, 497–502.

[Kot89] Kotta Ü. On the discretization of continuous-time linear-analytic systems (In Russian). *Proc. Estonian Acad. Sci. Phys. Math.*, 1989, v. 38, 222–224.

[Kot94] Kotta Ü. Discrete-time models of a nonlinear continuos-time system. *Proc. Estonian Acad. Sci. Phys. Math.*, 1994, v. 43, 64–78.

[Lev84] Levin S.A. Mathematical population biology. In: *Population Biology. Proc. of Symp. in Applied Mathematics.* American Mathematical Society, Providence, Rhode Island, 1984.

[May74] May R.M. Biological populations with nonoverlapping generations: stable points, stable cycles and chaos. *Science*, 1974, v. 186, 645–647.

[May78] May R. Mathematical aspects of the dynamics of animal populations. In: *Studies in Mathematical Biology. Part II. Populations and Communities.* Levin S.A. (Ed.). The Mathematical Association of America, 1978, 317–366.

[MNC85] Monaco S. and D.Normand-Cyrot. On the sampling of a linear analytic control system. *Proc. 24th IEEE CDC*, Fort Lauderdale, 1985, 1457–1462.

[MNC86a] Monaco S., and D.Normand-Cyrot. Approximation entree-sortie d'un systeme non lineaire continu par un systeme discret. *Lect. Notes in Contr. and Inf. Sciences*, 1986, v. 83, 354–367.

[MNC86b] Monaco S. and D.Normand-Cyrot. Nonlinear systems in discrete-time. *Lecture Notes in Control and Inf. Sci.*, 1986, v. 83, 411–430.

[MNC90] Monaco S., and D.Normand-Cyrot. A combinatorial approach of the nonlinear sampling problem. *Lect. Notes in Control and Inf. Sciences*, 1990, v. 144, 788–797.

[MSR57] Mickley H.S., T.K.Sherwood, and C.E.Reed. *Applied Mathematics in Chemical Engineering*. McGraw-Hill Book Company, New York, Toronto, London, 1957.

[Nij89] Nijmeijer H. On dynamic path decoupling and dynamic path controllability in economic systems. *J. of Economic Dynamics and Control*, 1989, v. 13, 21–39.

[NS90] Nijmeijer H., and A. van der Schaft. *Nonlinear Dynamical Control Systems*. Springer-Verlag, New York, 1990.

[Ost78] Oster G. The dynamics of nonlinear models with age structure. In: *Studies in Mathematical Biology. Part II. Populations and Communities*. Levin S.A. (Ed.) The Mathematical Association of America, 1978, 411–438.

[Pal83] Palm III W.J. *Modeling, Analysis and Control of Dynamic Systems*. John Wiley & Sons. New York, 1983.

[Pie77] Pielou E.C. *Mathematical Ecology*. John Wiley & Sons. New York-London-Sidney-Toronto, 1977.

[Sch85] Schnute J. A general theory for fishery modeling. *Lect. Notes in Biomathematics*, v. 61, 1985, 1–27.

[Swa84] Swan G.W. *Applications of Optimal Control Theory in Biomedicine*. Marcel Dekker, New York and Basel, 1984.

Part I

Control System Design for $(d_1, \ldots, d_p)$-forward Time-shift Right Invertible Systems

2. System Inversion. Special Case

The fundamental questions of the existence and construction of a right inverse system for the discrete-time nonlinear system of the form (1.11) are addressed in this chapter. Such a right inverse is intuitively understood to be a second discrete-time nonlinear system such that when the original system is applied in series with this right inverse, then its outputs are equal to the inputs of the right inverse system (see Fig. 2.1.).

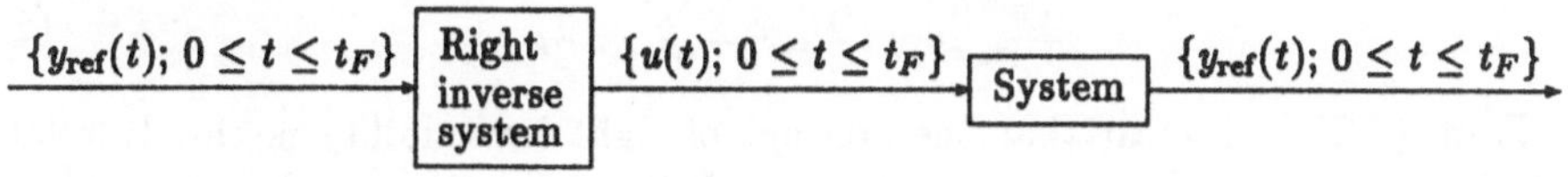

Fig. 2.1.

As such, the problem is of limited interest, since in a great number of cases no such inversion is possible. Greater generality is obtained by considering a notion of forward time-shift right inverse in which the input to the right inverse system is obtained by using the α-step forward time-shift operator δ^α on the reference signal: $\delta^\alpha y_{ref}(t) = y_{ref}(t+\alpha)$ (see Fig. 2.2.).

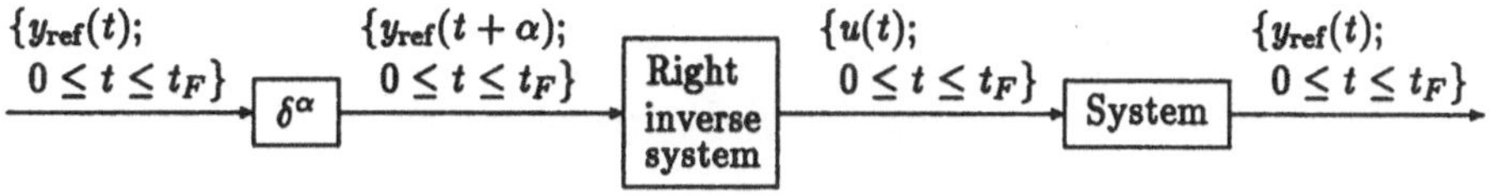

Fig. 2.2.

The determination of the smallest such α for a system (1.11) becomes a question of practical and theoretical importance In general, for multi-input multi-output systems it turns out that we must use different times the (one step) forward time-shift operator δ to different components of reference signal.

The results of this chapter allow determination of $d = \min \alpha$ (or $d_1 = \min \alpha_1, \ldots, d_p = \min \alpha_p$ for systems with p outputs) and a simple construction of $(d_1, \ldots, d_p)$-forward time-shift inverses.

2.1 The Concept of Right Invertibility

In this chapter we shall consider a discrete-time nonlinear control system S (without disturbances) of the form

$$\begin{aligned} x(t+1) &= f(x(t), u(t)), x(0) = x_0, \\ y(t) &= h(x(t), u(t)) \end{aligned} \tag{2.1}$$

where as in (1.11) x, u and y denote respectively the state, the control and the output.

It is natural to say that the system S is right invertible if there exists another system S_R^{-1}, called right inverse, such that the input-output map of the composition of S_R^{-1} and S is the identity map $\mathcal{I}_p$:

$$\Sigma_{(x_0, x_0^R)}^{S \circ S_R^{-1}} = \Sigma_{x_0}^{S} \circ \Sigma_{x_0^R}^{S_R^{-1}} = \mathcal{I}_p . \tag{2.2}$$

From (2.2) it is clear that the concept of right invertibility is closely related with the following problem in control theory. Is it possible to reproduce an arbitrary sequence $\mathbf{y}_{\text{ref}} \in \mathcal{Y}$ as an output of the given control system by manipulating the input? Or equivalently, is it possible to find a solution (not necessarily unique) of the equation

$$\Sigma_{x_0}^{S} \mathbf{u} = \mathbf{y}_{\text{ref}} \tag{2.3}$$

with respect to $\mathbf{u}$ for arbitrary $\mathbf{y}_{\text{ref}}$? Finding $\mathbf{u}$ from (2.3) can be viewed as looking for an inverse mapping $\left(\Sigma_{x_0}^{S}\right)^{-1}$. If the IO map of a given system is surjective, then the inverse mapping exists and the answer to the above question is obviously yes. It is also clear how to choose a control sequence $\mathbf{u}_{\text{ref}}$ that yields $\mathbf{y}_{\text{ref}}$

$$\mathbf{u}_{\text{ref}} = \left(\Sigma_{x_0}^{S}\right)^{-1} \mathbf{y}_{\text{ref}} . \tag{2.4}$$

So, the notions of right invertibility of a system and surjectivity of its input-output map are equivalent provided we may find another system S_R^{-1} with initial state x_0^R such that the inverse mapping of the original system is the IO map of this another system:

$$\Sigma_{x_0^R}^{S_R^{-1}} = \left(\Sigma_{x_0}^{S}\right)^{-1} . \tag{2.5}$$

If this is the case, the control sequence $\mathbf{u}_{\text{ref}}$ that yields $\mathbf{y}_{\text{ref}}$ may be found as the output of the inverse system S_R^{-1} if we take $\mathbf{y}_{\text{ref}}$ as the input of the inverse system

$$\mathbf{u}_{\text{ref}} = \Sigma_{x_0^R}^{S_R^{-1}} \mathbf{y}_{\text{ref}} . \tag{2.6}$$

Note that for the system of the form (2.1) the notions of right invertibility and surjectivity of its IO map are equivalent.

Since we have adopted a local viewpoint in this book, we are interested in local right invertibility (surjectivity) around an equilibrium point (x^0, u^0) of the system (2.1) and require (2.2) to hold only locally.

Next we shall give the formal definition of local right invertibility. In order to do this, let us denote by X^0 the open subset of X such that $\| x - x^0 \| < \gamma$ for some $\gamma > 0$. Let us also denote by $U^0 \subset U$ the set of the controls u, which are sufficiently close to u^0, that is $\| u - u^0 \| < \lambda$ for some $\lambda > 0$. Let us denote by $Y^0 \subset Y$ the set of the outputs y which are sufficiently close to $y^0 = h(x^0, u^0)$, that is $\| y - y^0 \| < \varepsilon$ for some $\varepsilon > 0$. Denote by $\mathcal{U}^0$ and $\mathcal{Y}^0$, the sets of the control sequences $\{u(t) \in U^0;\ 0 \le t \le t_F\}$ and the output sequences $\{y(t) \in Y^0;\ 0 \le t \le t_F\}$ respectively.

Definition 2.1 *The system (2.1) is called locally right invertible in a neighbourhood of its equilibrium point (x^0, u^0), if there exist sets $\mathcal{U}^0$ and $\mathcal{Y}^0$ such that given $x_0 \in X^0$, we are able to find for any sequence $\{y_{\mathrm{ref}}(t);\ 0 \le t \le t_F\} \in \mathcal{Y}^0$ a control sequence $\{u_{\mathrm{ref}}(t);\ 0 \le t \le t_F\} \in \mathcal{U}^0$ (not necessarily unique) such that*

$$y(t, x_0, u_{\mathrm{ref}}(0), \ldots, u_{\mathrm{ref}}(t)) = y_{\mathrm{ref}}(t)\,,\ 0 \le t \le t_F\,.$$

The above definition says that every sequence $\mathbf{y}_{\mathrm{ref}}$ whose elements are close enough to $y^0 = h(x^0, u^0)$ can be generated as the output signal.

Definition 2.2 *We call an equilibrium point (x^0, u^0) of the system (2.1) regular with respect to local right invertibility if the rank of the matrix $\partial h(x, u)/\partial u$ is constant around this point.*

It is well-known that the system (2.1) with $p \le m$ is locally right invertible in a neighbourhood of its regular equilibrium point (x^0, u^0) if and only if the rank of the matrix $\partial h(x, u)/\partial u$ is equal to p at the equilibrium point (x^0, u^0). This rank condition is certainly too restrictive for most systems and obviously useless for systems whose output map h does not depend explicitly on the control u. Such systems cannot be invertible in the sense of Definition 2.1 since the output y at $t = 0$ is not affected by the input and is completely defined by x_0, i.e. $y(0) = h(x_0)$. In general, the output may be defined completely by x_0 also at a few next time instances $t = 1, 2, \ldots, d-1$. Therefore, for those systems it is useless to require that all sequences could be locally reproducible and the best we can achieve is that all sequences could be locally reproducible beginning from certain time instant $t = d$.

We shall modify the definition of local right invertibility according to the above observations and introduce the notion of forward time-shift right invertibility.

2.2 The Concept of Forward Time-shift Right Invertibility

Definition 2.3 *The system (2.1) is called locally forward time-shift right invertible in a neighbourhood of its equilibrium point (x^0, u^0) if there exist integers $0 \le \alpha_1 \le$*

$\leq \alpha_2 \leq \ldots \leq \alpha_p$, a reordering of output components y_i, $i = 1,\ldots,p$, sets $\mathcal{U}^0$ and $\mathcal{Y}^0$ such that given $x_0 \in X^0$ we are able to find for any sequence $\{y_{\text{ref}}(t);\ 0 \leq t \leq t_F\} \in \mathcal{Y}^0$ a control sequence $\{u_{\text{ref}}(t);\ 0 \leq t \leq t_F\} \in \mathcal{U}^0$ yielding

$$y_i(t; x_0, u_{\text{ref}}(0), \ldots, u_{\text{ref}}(t)) = y_{\text{ref},i}(t)\,,\ \alpha_i \leq t \leq t_F\,,\ i = 1,\ldots,p\,.$$

Denote by Y_i^0 the set of output components y_i which are sufficiently close to $y_i^0 = h_i(x^0, u^0)$, that is $\| y_i - y_i^0 \| < \varepsilon_i$ for some $\varepsilon_i > 0$ and by $\mathcal{Y}_i^0$ the set of sequences $\{y_i(t) \in Y_i^0;\ 0 \leq t \leq t_F\}$. Then the above definition says that for the ith output component it is possible to reproduce locally all sequences $\mathbf{y}_{\text{ref},i}$ from $\mathcal{Y}_i^0$ beginning from time instant α_i. But forward time-shift right invertibility does not allow us to reproduce the first α_i terms in the arbitrary sequence $\{y_{\text{ref},i}(t);\ 0 \leq t \leq t_F\} \in \mathcal{Y}_i^0$.

Define the operator $\text{diag}\{\delta^{\alpha_1}, \ldots, \delta^{\alpha_p}\}$, acting on sequences $\mathbf{y}$ as follows

$$\text{diag}\{\delta^{\alpha_1}, \ldots, \delta^{\alpha_p}\}\mathbf{y} = (\delta^{\alpha_1}\mathbf{y}_1, \ldots, \delta^{\alpha_p}\mathbf{y}_p)^T$$

where for $i = 1,\ldots,p$

$$\delta^{\alpha_i}\mathbf{y}_i = \{y_i(t + \alpha_i);\ 0 \leq t \leq t_F\}\,,$$
$$\delta^{\alpha_i}y_i(t) = y_i(t + \alpha_i)\,,\ \text{for } 0 \leq t \leq t_F - \alpha_i\,,$$

and

$$\delta^{\alpha_i}y_i(t) = 0\,,\ \text{for } t_F - \alpha_i < t \leq t_F$$

since $y(t)$ is not defined for $t > t_F$.

Then the IO map of the forward time-shift right invertible system satisfies locally around the equilibrium point the equation

$$\text{diag}\{\delta^{\alpha_1}, \ldots, \delta^{\alpha_p}\} \circ \Sigma_{x_0}^S \circ \left(\Sigma_{x_0}^S\right)^{-1} = \text{diag}\{\delta^{\alpha_1}, \ldots, \delta^{\alpha_p}\} \circ \mathcal{I}_p\,. \tag{2.7}$$

A very important concept in treating system inversion from this generalized point of view, is the concept of delay orders. In the next section the set of system structural parameters, the so-called delay orders d_i, $i = 1,\ldots,p$, with respect to the control u of the system (2.1) are defined, one for each output component. The determination of these system structural parameters tells us immediately how many delays have been associated between the ith component y_i of the output and u and thus provides a lower bound on the number of output shifts required in an inverse system.

2.3 The Delay Orders with Respect to Control

With each component of the output y_i we can associate a delay order d_i (refered also in the literature as characteristic number or relative order) with respect to the control u in the following manner.

Denote the ith component of $h(x,u)$ by $h_i(x,u)$ and define $h_i^0(x,u) = h_i(x,u)$. Given an arbitrary state $x \in X$, and an arbitrary $u \in U$ we can compute for $i = 1, 2, \ldots, p$ the derivative

$$\frac{\partial}{\partial u} h_i^0(x,u).$$

From the analyticity of the system (2.1) it follows that either the vector $\partial h_i^0(x,u)/\partial u$ is nonzero for all (x,u) belonging to an open and dense subset O_i of $X \times U$ or this vector vanishes for all $(x,u) \in X \times U$. In the first case we define $d_i = 0$ whereas in the latter case we continue by observing that the function $h_i^0(x,u)$ does not depend on u and so we may write

$$h_i^0(x,u) = h_i^1(x)$$

for some analytic h_i^1 on X. Next we apply the one-step forward-shift operator δ to the equation

$$y_i(t) = h_i^1(x(t))$$

which gives

$$y_i(t+1) = h_i^1(f(x(t), u(t)))$$

and compute in an analogous fashion the derivate

$$\frac{\partial}{\partial u} h_i^1(f(x,u)).$$

If this vector is nonzero on an open and dense subset O_i of $X \times U$, we set $d_i = 1$; otherwise we continue with the function

$$h_i^2(x) = h_i^1(f(x,u)).$$

In this way the number d_i—if it exists—determines the inherent delay between the inputs and the ith output. Namely, the input $u(t)$ affects the ith output only after d_i steps, that is at the time instant $t + d_i$. In the case none of the iterated functions

$$h_i^{k+1}(x) = h_i^k(f(x,u)), \; k \geq 1,$$

depend on u, we define $d_i = \infty$. When $d_i = \infty$, the ith output evolves in time independently from the input sequence applied to the system (2.1). Thus, it seems reasonable to assume that all delay orders are finite.

Define by $\pi : X \times U \to X$ the projection along U on X.

Lemma 2.4 *Assume that $d_i < \infty$. Then the row vectors $\partial h_i^1(x)/\partial x, \ldots, \partial h_i^{d_i}(x)/\partial x$ are linearly independent on $\pi(O_i)$.*

Proof. For notational convenience we drop the index i. By the definition of delay order, using

$$\frac{\partial h^j}{\partial x}\left(\frac{\partial f}{\partial x}\right)^k \frac{\partial f}{\partial u} = \frac{\partial h^{j+k}}{\partial x}\frac{\partial f}{\partial u} = \frac{\partial}{\partial u} h^{j+k}(f(x,u)), \; j + k \leq d,$$

we have

$$\frac{\partial h^j}{\partial x}\left(\frac{\partial f}{\partial x}\right)^k \frac{\partial f}{\partial u} = \begin{cases} 0\,, \text{ if } j+k<d \\ \frac{\partial}{\partial u} h^d(f(x,u)) \neq 0\,, \text{ if } j+k=d, \text{ and } (x,u)\in O\,. \end{cases}$$

The above condition shows that the matrix

$$\begin{bmatrix} \frac{\partial h^1}{\partial x} \\ \vdots \\ \frac{\partial h^d}{\partial x} \end{bmatrix} \left[\frac{\partial f}{\partial u}, \frac{\partial f}{\partial x}\frac{\partial f}{\partial u}, \ldots, \left(\frac{\partial f}{\partial x}\right)^d \frac{\partial f}{\partial u}\right] = \begin{bmatrix} 0 & \ldots & \frac{\partial}{\partial u} h^d(f(x,u)) \\ 0 & \ldots & * \\ \ldots & \ldots & \ldots \\ \frac{\partial}{\partial u} h^d(f(x,u)) & \ldots & * \end{bmatrix}$$

has rank d in O and thus, the row vectors $\partial h^1(x)/\partial x, \ldots, \partial h^d(x)/\partial x$ are linearly independent on $\pi(O)$. ∎

The following result gives an upper bound on each finite delay order.

Lemma 2.5 *Each finite delay order d_i, $i = 1, \ldots, p$, satisfies the inequality $d_i \leq n$.*

Proof. For notational convenience we drop the index i. If the delay order d is finite but greater than the dimension of state vector, $d > n$, then by Lemma 2.4 the $n+1$ rows of the $(n+1) \times n$-dimensional matrix

$$\frac{\partial}{\partial x} H(x) = \frac{\partial}{\partial x} \begin{bmatrix} h^1(x) \\ \ldots \\ h^n(x) \\ h^{n+1}(x) \end{bmatrix}.$$

would be linearly independent on O which is clearly impossible. The contradiction proves the lemma. ∎

Lemma 2.6 *For arbitrary feedback, the delay orders of the original system are less than or equal to the corresponding delay orders of the feedback modified system, $d_i \leq \tilde{d}_i$, $i = 1, \ldots, p$. In case of regular static state feedback the equality holds.*

Proof. At first we show that

$$\tilde{h}_i^k(x) = h_i^k(x)$$

for all $1 \leq k \leq d_i$ and all $x \in O_i$. This fact can be easily proved by induction. It is true for $k = 1$ and, if true for some $1 \leq k < d_i$, then

$$\tilde{h}_i^{k+1}(x) = \tilde{h}_i^k(\tilde{f}(x,v)) = h_i^k(f(x,\alpha(x,v))) = h_i^{k+1}(x)\,.$$

This proves $d_i \leq \tilde{d}_i$.

The equality of the delay orders under regular static state feedback follows from

$$\frac{\partial}{\partial v}\tilde{h}_i^{\tilde{d}_i}(\tilde{f}(x,v)) = \frac{\partial}{\partial u} h_i^{d_i}(f(x,u)) \cdot \frac{\partial}{\partial v}\alpha(x,v)\,.$$

∎

2.4 The Definition of $(d_1,\ldots,d_p)$-forward Time-shift Right Invertibility

From the definition of delay orders d_i with respect to the control u it is clear that the initial part of every reproducible sequence must satisfy the following conditions

$$\begin{aligned} y_i(0) &= h_i^1(x_0) \\ y_i(1) &= h_i^2(x_0) \\ &\cdots \\ y_i(d_i-1) &= h_i^{d_i}(x_0) \end{aligned} \tag{2.8}$$

and that only for $y_i(d_i)$ arise the possibility to change it arbitrarily.

The notion of $(d_1,\ldots,d_p)$-forward time-shift right invertibility can be obtained as a special case of Definition 2.3.

Definition 2.7 *The system (2.1) is called locally $(d_1,\ldots,d_p)$-forward time-shift ($(d_1,\ldots,d_p)$-FTS) right invertible in a neighbourhood of its equilibrium point (x^0,u^0) if there exist sets $\mathcal{U}^0$ and $\mathcal{Y}^0$ such that given $x_0 \in X^0$ we are able to find for any sequence $\{y_{\mathrm{ref}}(t)\,;0\le t\le t_F\}\in\mathcal{Y}^0$ a sequence of controls $\{u_{\mathrm{ref}}(t)\,;\ 0\le t\le t_F\}\in\mathcal{U}^0$ yielding*

$$y_i(t;x_0,u_{\mathrm{ref}}(0),\ldots,u_{\mathrm{ref}}(t)) = y_{\mathrm{ref},i}(t)\,,\ d_i\le t\le t_F\,,\ i=1,\ldots,p\,.$$

The above definition says that for ith output it is possible to reproduce all sequences from $\mathcal{Y}_i^0$ beginning from time instant d_i. But $(d_1,\ldots,d_p)$-FTS right invertibility does not allow us to reproduce the first d_i terms in the arbitrary sequence $\{y_{\mathrm{ref},i}(t);\ 0\le t\le t_F\}$ from $\mathcal{Y}_i^0$.

The input-output map of the $(d_1,\ldots,d_p)$-FTS right inverse system S_R^{-1} satisfies locally around the equilibrium point the equation

$$\mathrm{diag}\{\delta^{d_1},\ldots,\delta^{d_p}\}\circ\Sigma_{x_0}^S\circ\Sigma_{x_0^R}^{S_R^{-1}} = \mathrm{diag}\{\delta^{d_1},\ldots,\delta^{d_p}\}\circ\mathcal{I}_p\,, \tag{2.9}$$

where by δ^d is denoted the d-step forward time-shift operator and by $\mathcal{I}_p$ the identity map.

Note that the adjective local in this definition is related to the fact that the reproducibility property for a nonlinear system in general holds only in the neighbourhood $X^0\times U^0$ of the equilibrium point (x^0,u^0) and for reference sequences $\{y_{\mathrm{ref}}(t);\ 0\le t\le t_F\}$ from restricted set of output sequences $\mathcal{Y}^0$.

2.5 Necessary and Sufficient Conditions for $(d_1,\ldots,d_p)$-forward Time-shift Right Invertibility

To be useful in a state-space approach, the conditions for invertibility should be phrased directly in terms of the functions f and h. Such a condition is developed in this section.

Rather than first looking for conditions under which S is right invertible, we shall directly attack the problem of obtaining a representation S_R^{-1} for the inverse mapping $(\Sigma_{x0}^S)^{-1}$ when it exists.

The basic idea is to apply the one step forward time-shift operator δ on those output components which do not explicitly depend on the input. Assuming that each delay order d_i is finite, we modify the output equation of the system by repeatedly operating on each of the scalar output equations the forward time-shift operators so as to obtain a system of equations each of which depends explicitly on the control $u(t)$. From the definition of delay orders d_i, $i = 1, \dots, p$, we obtain

$$\delta^{d_1} y_1(t) = y_1(t + d_1) = h_1^{d_1}(f(x(t), u(t)))$$
$$\dots$$
$$\delta^{d_p} y_p(t) = y_p(t + d_p) = h_p^{d_p}(f(x(t), u(t)))$$

(where $h_i^0(f(x,u)) = h_i(x,u)$ by definition), or in the vector form

$$\begin{bmatrix} y_1(t+d_1) \\ \vdots \\ y_p(t+d_p) \end{bmatrix} = A(x(t), u(t)) \,. \tag{2.10}$$

We introduce the so-called decoupling matrix $K(x,u)$ in the following way

$$K(x,u) = \frac{\partial}{\partial u} A(x,u) = \frac{\partial}{\partial u} \begin{bmatrix} h_1^{d_1}(f(x,u)) \\ \dots \\ h_p^{d_p}(f(x,u)) \end{bmatrix} .$$

From the definition of the d_i's the rows of the matrix $K(x,u)$ are nonzero functions on an open and dense subset $O = O_1 \cap O_2 \cap \dots \cap O_p$ of $X \times U$. It is obvious that the rank of $K(x,u)$ is, in general, state and control dependent. To ensure smoothness of the solution of (2.10) with respect to $u(t)$ we have to assume that $K(x,u)$ has a constant rank. This assumption is formalized in the notion of regularity of an equilibrium point.

Definition 2.8 *We call the equilibrium point (x^0, u^0) of the system (2.1) regular with respect to $(d_1, \dots, d_p)$-forward time-shift right invertibility, if the rank of the decoupling matrix $K(x,u)$ of the system (2.1) is constant around (x^0, u^0).*

Theorem 2.9 *Assume that for the system (2.1) $d_i \le n$, $i = 1, \dots, p$. Then the system (2.1) is locally $(d_1, \dots, d_p)$-forward time-shift right invertible around a regular equilibrium point (x^0, u^0) if and only if rank $K(x^0, u^0) = p$.*

Proof. Sufficiency. Consider the system of equations (2.10). By the definition of the equilibrium point we have $y^0 = A(x^0, u^0)$. Observe that the Jacobian matrix of the right hand side of (2.10) with respect to the control u equals to the decoupling matrix $K(x,u)$. By the assumption of the theorem the rank of the decoupling matrix $K(x,u)$ is equal to p at the equilibrium point (x^0, u^0). So, we may apply the Implicit Function Theorem in order to solve the system of equations (2.10) with respect to the control

u. After a possible reordering of the control components we may assume that the Jacobian matrix of the right hand side of (2.10) with respect to $u^1 = (u_1, \ldots, u_p)^T$ around the point (x^0, u^0) has full row rank p. Therefore, equation (2.10) can be solved for $u^1(t)$ uniquely around (x^0, u^0). Define $u^2 = (u_{p+1}, \ldots, u_m)^T$.

The Implicit Function Theorem says that in some (possible small) neighbourhood $\bar{X}^0 \times \bar{U}^0 \times \bar{Y}^0$ of (x^0, u^0, y^0) there exists an analytic function φ of variables $x(t)$, $y_1(t + d_1), \ldots, y_p(t + d_p)$, and $u^2(t)$ i.e.

$$u^1(t) = \varphi(x(t), y_1(t + d_1), \ldots, y_p(t + d_p), u^2(t)) \tag{2.11}$$

which is such that

$$\varphi(x^0, y^0, u^{20}) = u^{10}$$

and

$$[y_1(t + d_1), \ldots, y_p(t + d_p)]^T \equiv A(x(t), \varphi(x(t), y_1(t + d_1), \ldots, y_p(t + d_p), u^2(t)), u^2(t)).$$

Notice that the above identity is lost if we leave the neighbourhoods $\bar{X}^0 \times \bar{Y}^0 \times \bar{U}^{20}$ or $\bar{U}^{10}$.

Necessity. Suppose that the system (2.1) is locally $(d_1, \ldots, d_p)$-FTS right invertible around its regular equilibrium point (x^0, u^0). This implies, in particular, that at time instant $t = d_i$ at the ith output y_i of the system (2.1) we can reproduce by suitable choice of $u(0) = u_{\text{ref}}$ arbitrary $y_{\text{ref},i}$ sufficiently close to y_i^0, that is the following holds

$$h_i^{d_i}(f(x_0, u_{\text{ref}})) = y_{\text{ref},i}\,, \; i = 1, \ldots, p.$$

Assume that rank $K(x^0, u^0) = k < p$. As by regularity of (x^0, u^0), k is constant in some neighbourhood $\bar{X}^0 \times \bar{U}^0$ of (x^0, u^0), the rank $K(x, u)$ in $X^0 \times U^0$ is less than p. This implies that the functions $h_i^{d_i}(f(x_0, u_{\text{ref}}))$, $i = 1, \ldots, p$ of u_{ref} are functionally dependent, that is there exists the map $R(\cdot)$ such that

$$R(h_i^{d_1} \circ f, \ldots, h_p^{d_p} \circ f) = R(y_{\text{ref},1}, \ldots, y_{\text{ref},p}) = 0\,.$$

The last equality means that y_{ref} is not arbitrary but satisfies the equation $R(y_{\text{ref},1}, \ldots, y_{\text{ref},p}) = 0$ which gives contradiction. ■

Remark. Clearly, rank $K(x^0, u^0) = p$ requires $m \geq p$. So, $p \leq m$ is always a necessary condition for system to have a $(d_1, \ldots, d_p)$-FTS right inverse, that is the system must have least as many inputs as outputs.

Remark. We should like to stress that the assumption of the regularity of the equilibrium point (x^0, u^0) in Theorem 2.9 is extremely vital. If the point (x^0, u^0) is not regular, that is around the point (x^0, u^0) the rank of the decoupling matrix $K(x, u)$ is not necessarily constant, then the condition $K(x^0, u^0) = p$ is not necessary for $(d_1, \ldots, d_p)$-FTS right invertibility. The illustration of this phenomenon is given in the following simple example:

$$\begin{aligned} x(t+1) &= u^3(t) \\ y(t) &= x(t)\,. \end{aligned}$$

We have

$$K(x,u) = \frac{\partial}{\partial u} A(x,u) = 3u^2 .$$

At the equilibrium point $x^0 = 0$, $u^0 = 0$ the rank of $K(x,u)$ is equal to 0 which is less than $p = 1$. Still, the arbitrary sequences are reproducible by the choice of control

$$u(t) = \sqrt[3]{y(t+1)} .$$

The reason is that the point $x^0 = 0$, $u^0 = 0$ is not a regular equilibrium point. The rank of the matrix $K(x,u)$ is equal to 1 at all points $u \neq 0$.

2.6 Construction of $(d_1, \ldots, d_p)$-forward Time-shift Right Inverse System

The system S_R^{-1} is said to be a $(d_1, \ldots, d_p)$-forward time-shift right inverse for S, if its IO map satisfies equation (2.9).

We are now going to derive the equations of S_R^{-1}. Apply to the each scalar output equation of the system (2.1) the one step forward time-shift operator δ until if becomes explicitly dependent on the control $u(t)$, that is for the ith output equation d_i times. Doing this we obtain equation (2.10). To get the equations of right inverse, we have to be able to solve the system of equations (2.1), (2.10) with respect to $u(t)$ and $x(t+1)$ in terms of $x(t)$ and $y_1(t+d_1), \ldots, y_p(t+d_p)$. The solution of (2.10) is given by (2.11) with arbitrary $u^2(t) \in \bar{U}^{20}$. Equation (2.11) defines the required control for the given system (which yields the reference output) in terms of state and reference output. We can take this equation as the output equation of a right inverse system. To obtain the dynamic part of the right inverse, we must substitute (2.11) into (2.1)

$$x(t+1) = f(x(t), \varphi(x(t), y_1(t+d_1), \ldots, y_p(t+d_p), u^2(t)), u^2(t)) . \tag{2.12}$$

So, the equations of S_R^{-1} are the following

$$\begin{aligned} x^R(t+1) &= f(x^R(t), \varphi(x^R(t), y_1(t+d_1), \ldots, y_p(t+d_p), \lambda(t)), \lambda(t)) \\ u^1(t) &= \varphi(x^R(t), y_1(t+d_1), \ldots, y_p(t+d_p), \lambda(t)) \\ u^2(t) &= \lambda(t) , \end{aligned} \tag{2.13}$$

where by $\lambda(t)$ is denoted the free parameter.

Remark. Note that the states $x(t)$ of the system (2.1) under the state feedback (2.11) and the states of the $(d_1, \ldots, d_p)$-FTS right inverse system (2.13) coincide, provided the initial states of these systems are equal, i.e. if $x(0) = x^R(0)$.

From the equations (2.13) it is clear that the state space systems of the form (2.1) are not closed under system inversion: the inverse system in general depends on the future values of the outputs of the original system.

There exist no right inverse for a strictly causal system S such that the input to the original system can be computed as the output of the inverse system without using the future values of the reference signal, or equivalently, without using forward shift operators on reference signal. To be more precise we must apply the d_i-step forward time-shift operator for ith component of reference signal.

Actually, the output of the right inverse system (i.e. the control of the original system) at time instant t will depend on the ith component of the reference output at time instant $t+d_i$. As $\mathbf{y}_{\text{ref}}$ is generated by the designer, in the actual control law design this implies that a change of reference signal must be preplanned some time steps ahead, which is often a realistic assumption. When it is possible, the right inverse system can be realized.

If the reference signal can be generated from a model M, the need for future values of model inputs is avoided under conditions that the composite system $S_R^{-1} \circ M$ contains no forward time-shift operators.

The disadvantage of the definition (2.9) lies in the fact that S_R^{-1}, defined by it, is not, in general, a causal system (see equations (2.13)), and therefore, cannot be realized via state equations. One can decompose the noncausal system S_R^{-1} into two systems, a causal system $\bar{S}_R^{-1}$ and a system which transforms signals by applying on them forward time-shift operator $\operatorname{diag}\{\delta^{d_1},\ldots,\delta^{d_p}\}$. Hence, one can express the input-output map of S_R^{-1}, that is $\Sigma_{x_0^R}^{S_R^{-1}}$, in the following form

$$\Sigma_{x_0^R}^{S_R^{-1}} = \Sigma_{\bar{x}_0^R}^{\bar{S}_R^{-1}} \circ \operatorname{diag}\{\delta^{d_1},\ldots,\delta^{d_p}\} .$$

Then, one can define a causal $(d_1,\ldots,d_p)$-forward time-shift right inverse $\bar{S}_R^{-1}$ of S as such which satisfies the equation

$$\operatorname{diag}\{\delta^{d_1},\ldots,\delta^{d_p}\} \circ \Sigma_{x_0}^{S} \circ \Sigma_{\bar{x}_0^R}^{\bar{S}_R^{-1}} \circ \operatorname{diag}\{\delta^{d_1},\ldots,\delta^{d_p}\} = \operatorname{diag}\{\delta^{d_1},\ldots,\delta^{d_p}\} \circ \mathcal{I}_p . \quad (2.14)$$

Note that $\bar{S}_R^{-1}$ can be realized via the state equations

$$\begin{aligned} \bar{x}^R(t+1) &= f(\bar{x}^R(t),\, \varphi(\bar{x}^R(t),\, u^R(t), \lambda(t)), \lambda(t)) \\ y^R(t) &= \begin{bmatrix} \varphi(\bar{x}^R(t),\, u^R(t), \lambda(t)) \\ \lambda(t) \end{bmatrix} , \end{aligned} \quad (2.15)$$

where by u^R and y^R are denoted the input and output of $\bar{S}_R^{-1}$ respectively.

Remark. The states $x(t)$ of the system (2.1) under the state feedback $u^1(t) = \varphi(x(t), u^R(t), \lambda(t)), u^2(t) = \lambda(t)$ and the states of the causal $(d_1,\ldots,d_p)$-FTS right inverse system (2.15) under input $u^R(t)$ coincide, provided the initial states of these systems are equal, i.e. $x(0) = \bar{x}^R(0)$.

2.7 An Approximate $(d_1,\ldots,d_p)$-forward Time-shift Right Inverse System

The Implicit Function Theorem says that equation (2.10) can be solved for $u(t)$ but does not indicate how to find the solution. Assume for simplicity that $m = p$. We recall here a result due to Gröbner for the inversion of a family of analytic functions

$$A_i(u_1,\dots,u_p) = y_i\,,\ i = 1,\dots,p\,.$$

Assuming that $\det\{\partial A_i(\cdot)/\partial u_j\} \neq 0$ at a fixed point u^*, the inverse functions $u_i\,,\ i = 1,\dots,p$ can be explicitly expressed by means of Lie series which are absolutely and uniformly convergent in a neighbourhood of (u^*, y^*) such that $y_i^* = A_i(u^*)$:

$$u = \sum_{\nu_1=0,\dots,\nu_p=0}^{\infty} \frac{1}{\nu_1!\dots\nu_p!} D_1^{\nu_1}\dots D_p^{\nu_p} u_i\Big|_{u=u^*} (y_1 - y_1^*)^{\nu_1}\dots(y_p - y_p^*)^{\nu_p}$$

where $D_k\,,\ k = 1,\dots,p$ are the following commuting linear differential operators

$$D_k = \sum_{i=1}^{p} \hat{A}_{ik}(u)\frac{\partial}{\partial u_i}\,,\quad \{\hat{A}_{ik}\} = (\partial A_i(\cdot)/\partial u_j)^{-1}\,.$$

So, denoting by $\hat{A}_{ik}(x,u)$ the elements of the matrix $\partial A(x,u)/\partial u$ and by $D_j(x,u)\,,\ j = 1,\dots,p$ the differential operators $\Sigma_{i=1}^{p}\hat{A}_{ij}(x,u)\frac{\partial}{\partial u_i}$, the solution of equation (2.10) with respect to the components u_i of the control vector u in the neighbourhood of (x^0,u^0,y^0) can be expressed via the series

$$\begin{aligned} u_i(t) = &\sum_{\nu_1=0,\dots,\nu_p=0}^{\infty} \frac{1}{\nu_1!\dots\nu_p!} D_1^{\nu_1}(x(t),u)\circ\dots \\ &\dots D_p^{\nu_p}(x(t),u)u_i\Big|_{u=u^0}[(y_1(t+d_1) - y_1^0]^{\nu_1}\dots[(y_p(t+d_p) - y_p^0]^{\nu_p}\,, \\ i = 1,\dots,p\,. \end{aligned} \tag{2.16}$$

The exact solution (2.16) of (2.10) is defined via infinite series and in reality, to compute the solution, one must confine oneself with finite number of terms in this series. In that way we reach the approximate solution of (2.10) and the notion of approximate $(d_1,\dots,d_p)$-FTS right inverse system, if we substitute (2.16), truncated at $\nu_1 + \dots + \nu_p = t$, into (2.1).

2.8 The Reduced Order Inverse System. State-free Inverse System

The order of the inverse system (2.13) (or (2.15)), discussed in Section 2.6, is the same as that of the original system. It will now be shown that, provided the system is $(d_1,\dots d_p)$-forward time-shift right invertible, an inverse system of order $n - \Sigma_{i=1}^{p} d_i$ can always be obtained if we utilize $y_i(t+j)\,,\ 0 \le j \le d_i - 1\,,\ 1 \le i \le p$ as the additional inputs of an inverse system. (Note that the inputs of the right inverse system of order n are $y_1(t+d_1),\dots,y_p(t+d_p)$). Thus, if $d_1 + \dots + d_p$ is close to n, a relatively low order inverse system can be found.

In the sequel we need the following Lemma.

Lemma 2.10 *Suppose that the system (2.1) is locally $(d_1, \ldots, d_p)$-forward time-shift right invertible in the neighbourhood $X^0 \times U^0$ of its regular equilibrium point (x^0, u^0). Then the functions $h_i^j(x), 1 \le j \le d_i, 1 \le i \le p$ are functionally independent on $\pi(X^0 \times U^0)$.*

Proof. The proof is similar to the proof of Lemma 2.4. Define $d = \max\{d_1, \ldots, d_p\}$ and

$$H(x) = \begin{bmatrix} h_1^1(x) \\ \vdots \\ h_1^{d_1}(x) \\ \vdots \\ h_p^1(x) \\ \vdots \\ h_p^{d_p}(x) \end{bmatrix}.$$

Consider the matrices $P(x) = \partial H(x)/\partial x$ and

$$Q(x,u) = \left[\frac{\partial f}{\partial u}, \frac{\partial f}{\partial x}\frac{\partial f}{\partial u}, \ldots, \left(\frac{\partial f}{\partial x}\right)^d \frac{\partial f}{\partial u}\right].$$

Using

$$\frac{\partial h_i^j}{\partial x}\left(\frac{\partial f}{\partial x}\right)^k \frac{\partial f}{\partial u} = \frac{\partial h_i^{j+k}}{\partial x}\frac{\partial f}{\partial u} = \frac{\partial}{\partial u} h_i^{j+k}(f(x,u)), \; j+k \le d_i$$

and the definition of delay order, it is easy to see that the matrix

$$P(x) \cdot Q(x,u),$$

after possibly a reordering of the rows, exhibits a block triangular structure in which the diagonal blocks consist of rows of the decoupling matrix $K(x,u)$. This shows the linear independence of the rows of $\partial H(x)/\partial x$ in $\pi(X^0 \times U^0)$ which completes the proof. ∎

Remark. If all $d_i < \infty$, an immediate implication of the above lemma is that the sum of relative degrees of an nth order system is always equal or less than n, because otherwise there would be more than n linearly independent vectors in an n-dimensional space.

Our approach to construct a reduced-order inverse is to construct first a non-reduced inverse of order n and then reduce its dimension by expressing some components of the state via the outputs $y_i(t+j), \; 0 \le j \le d_i - 1, \; 1 \le i \le p$ and the remaining components of state.

The following equalities result from the definition of delay orders d_i of the system (2.1) (See Section 2.3):

$$y_i(t+j) = h_i^{j+1}(x(t)), \; j = 0, \ldots, d_i - 1, \; i = 1, \ldots, p. \tag{2.17}$$

In case of $(d_1,\dots,d_p)$-FTS invertible systems by Lemma 2.10 the functions $h_i^j(x)$, $1 \le j \le d_i$, $1 \le i \le p$ are functionally independent on $\pi(X^0 \times U^0)$. After a possible reordering of the state components we may assume that

$$\operatorname{rank} \frac{\partial}{\partial(x_1,\dots,x_\mu)} H(x) = \mu .$$

Therefore, equations (2.17) can be solved by Implicit Function Theorem with respect to $x^1(t) = (x_1(t),\dots,x_\mu(t))$

$$x^1(t) = \Phi(\{y_i(t+j)\,,\ j=0,\dots,d_i-1\,,\ i=1,\dots,p\}\,,\ x^2(t))\,,$$

where $x^2(t) = (x_{\mu+1}(t),\dots,x_n(t))$. Substituting $x^1(t)$ into (2.11) we have

$$\begin{aligned} u^1(t) = \varphi(&\Phi(\{y_i(t+j), j=0,\dots,d_i-1, i=1,\dots,p\}\,, x^2(t)), x^2(t),\\ &y_1(t+d_1),\dots,y_p(t+d_p), u^2(t)) \end{aligned}$$

and consequently, the dynamics of the right inverse system (2.13) of order n can be replaced by equations of order $n-\mu$:

$$\begin{aligned} x^{R2}(t+1) = {}& f^2(\Phi(\{y_i(t+j), 1\le i\le p, 0\le j\le d_i-1\}, x^{R2}(t)), x^{R2}(t),\\ &\varphi(\Phi(\{y_i(t+j), 1\le i\le p, 0\le j\le d_i-1\}, x^{R2}(t)), x^{R2}(t),\\ &y_1(t+d_1),\dots,y_p(t+d_p), \lambda(t)), \lambda(t)) :=\\ :={}& \hat f(x^{R2}(t), \{y_i(t+j)\,,\ j=0,\dots,d_i\,,\ i=1,\dots,p\}, \lambda(t)) . \end{aligned} \tag{2.18}$$

Note that the number of inputs of (2.18) is $p+\sum_{i=1}^p d_i$ which is of course greater than p – the number of inputs of (2.13).

If $\dim x^2 = 0$, that is if $\sum_{i=1}^p d_i = n$, the reduced order right inverse is said to be „state-free". In that case the input, which produces the reference signal, can be computed from the knowledge of the reference signal $\{y_{ref}(t)\,;\ 0\le t\le t_F\}$ only:

$$\begin{aligned} u^1(t) = {}&\varphi(\Phi(\{y_i(t+j), j=0,\dots,d_i-1, i=1,\dots,p\}), y_1(t+d_1),\dots\\ &\dots, y_p(t+d_p), u^2(t)) \triangleq \bar\varphi(\{y_i(t+j), j=0,\dots,d_i, i=1,\dots,p\}, u^2(t)) . \end{aligned}$$

2.9 Examples

In this section we show on two examples how to construct the $(d_1,\dots,d_p)$-FTS right inverse system.

Example 2.11 Bilinear model of neutron kinetics can be described by the following equations

$$\begin{aligned} x_1(t+1) &= (1-a_1)x_1(t) + a_1x_2(t) + b_1x_1(t)u(t)\\ x_2(t+1) &= a_2x_1(t) + (1-a_2)x_2(t)\\ y(t) &= x_1(t) . \end{aligned} \tag{2.19}$$

At first we shall find the delay order. Compute

$$\frac{\partial}{\partial u(t)}h(x(t+1)) = \frac{\partial}{\partial u(t)}x(t+1) = b_1 x_1(t)$$

which is different from zero if $x_1(t) \neq 0$. So, $d = 1$.

Now we are going to construct the d-forward time-shift right inverse system for neutron kinetics model. Following the theory in Section 2.6 we apply to the output equation of the system (2.19) the forward shift operator δ until it becomes explicitly dependent on the control $u(t)$. In our example we need to apply the operator δ only once. Doing this we reach the equation

$$y(t+1) = (1-a_1)x_1(t) + a_1 x_2(t) + b_1 x(t)u(t)\,. \tag{2.20}$$

To get the output equation of the system S_R^{-1} we must solve equation (2.20) for $u(t)$:

$$u(t) = -\frac{(1-a_1)x_1(t) + a_1 x_2(t)}{b_1 x_1(t)} + \frac{1}{b_1 x_1(t)}y(t+1)\,. \tag{2.21}$$

To obtain the dynamic part of the S_R^{-1} we must substitute (2.21) into (2.19):

$$\begin{aligned} x_1(t+1) &= -y(t+1) \\ x_2(t+1) &= a_2 x_1(t) + (1-a_2)x_2(t)\,. \end{aligned} \tag{2.22}$$

From the equations (2.21), (2.22) it is easy to obtain the d-forward time-shift right inverse system $\bar{S}_R^{-1}$

$$\begin{aligned} x_1(t+1) &= -u^R(t) \\ x_2(t+1) &= a_2 x_1(t) + (1-a_2)x_2(t) \\ y^R(t) &= -\frac{(1-a_1)x_1(t) + a_1 x_2(t)}{b_1 x_1(t)} + \frac{1}{b_1 x_1(t)}u^R(t)\,. \end{aligned} \tag{2.23}$$

To construct the reduced-order right inverse system we replace in equations (2.21), (2.22) $x_1(t)$ by $y(t)$:

$$\begin{aligned} x_2(t+1) &= a_2 y(t) + (1-a_2)x_2(t) \\ u(t) &= -\frac{(1-a_1)y(t) + a_1 x_2(t)}{b_1 y(t)} + \frac{1}{b_1 y(t)}y(t+1)\,. \end{aligned}$$

Example 2.12 DC/AC convertor.

Consider the system

$$\begin{aligned} x_1(t+1) &= -4x_1(t) - \frac{x_2^2(t)}{x_1(t)} + 5u_1(t) \\ x_2(t+1) &= -6x_2(t) - \frac{x_2^3(t)}{x_1^2(t)} + 5\frac{x_2(t)}{x_1(t)}u_1(t) + 2x_1(t)u_2(t) \\ y_1(t) &= x_1(t)\,,\; y_2(t) = x_2(t)\,. \end{aligned}$$

At first, let us find the delay orders d_1 and d_2. Compute

$$\frac{\partial}{\partial u(t)}h^1(x(t+1)) = \frac{\partial}{\partial u(t)}x_1(t+1) = [5,0] \neq 0\,,$$
$$\frac{\partial}{\partial u(t)}h^2(x(t+1)) = \frac{\partial}{\partial u(t)}x_2(t+1) = \left[5\frac{x_2(t)}{x_1(t)},\ 2x_1(t)\right] \neq 0\,,$$

if $x_1(t) \neq 0$.

So, $d_1 = d_2 = 1$.

We are now going to check if this system is $(d_1, \ldots, d_p)$-forward time-shift right invertible. For this purpose, we have to construct the decoupling matrix $K(x,u)$ and check if the rank of this matrix equals to the number of the outputs of the system. The rank of the decoupling matrix

$$K(x,u) = \begin{bmatrix} 5 & 0 \\ \dfrac{5x_2}{x_1} & 2x_1 \end{bmatrix}$$

is equal to 2, if $x_1 \neq 0$.

Now, let us find the equations of $(d_1, \ldots, d_p)$-forward time-shift right inverse system $\bar{S}_R^{-1}$. First we derive the equations of S_R^{-1}. For that we need to solve the system of equations

$$y_1(t+1) = -4x_1(t) - \frac{x_2^2(t)}{x_1(t)} + 5u_1(t)$$
$$y_2(t+1) = -6x_2(t) - \frac{x_2^3(t)}{x_1^2(t)} + 5\frac{x_2(t)}{x_1(t)}u_1(t) + 2x_1(t)u_2(t)$$

with respect to $u_1(t)$ and $u_2(t)$:

$$u_1(t) = \frac{1}{5}\left\{y_1(t+1) + 4x_1(t) + \frac{x_2^2(t)}{x_1(t)}\right\}$$
$$u_2(t) = -\frac{x_2(t)}{2x_1^2(t)}\left\{y_1(t+1) + 4x_1(t) + \frac{x_2^2(t)}{x_1(t)}\right\} +$$
$$+\frac{1}{2x_1(t)}\left\{y_2(t+1) + 6x_2(t) + \frac{x_2^3(t)}{2x_1^2(t)}\right\}.$$

These are the output equations of the system S_R^{-1}. To obtain the output equations of $\bar{S}_R^{-1}$, we must replace $y_1(t+1)$ by $u_1^R(t)$ and $y_2(t+1)$ by $u_2^R(t)$. At last, to obtain the state equations of right inverse, we must substitute so found $u_1(t)$ and $u_2(t)$ into state equations of the original system

$$x_1(t+1) = -4x_1(t) - \frac{x_2^2(t)}{x_1(t)} + u_1^R(t) + 4x_1(t) + \frac{x_2^2(t)}{x_1(t)} = u_1^R(t)$$
$$x_2(t+1) = -6x_2(t) - \frac{x_2^3(t)}{x_1^2(t)} + \frac{x_2(t)}{x_1(t)}\left\{u_1^R(t) + 4x_1(t) + \frac{x_2^2(t)}{x_1(t)}\right\} -$$
$$-x_1(t)\frac{x_2(t)}{x_1^2(t)}\left\{u_1^R(t) + 4x_1(t) + \frac{x_2^2(t)}{x_1(t)}\right\} +$$
$$+x_1(t)\frac{1}{x_1(t)}\left\{u_2^R(t) + 6x_2(t) + \frac{x_2^3(t)}{x_1^2(t)}\right\} = u_2^R(t).$$

Notes and References

The delay orders were first introduced in the discrete-time nonlinear context by Monaco and Normand-Cyrot [MNC83b] for the very restricted subclass of linear-analytic systems. The definition of delay orders used here is taken from [Nij87]. Note that sometimes one defines the delay order as the minimal number of d for which $y(t+d+1)$ (and not $y(t+d)$ as in our definition) depends explicitly on $u(t)$ [MNC83b], [Kot86a], [Kot90]. As we do allow the output function in (2.1) explicitly depend on the control, this slight change in the definition will avoid possible introduction of negative delay order. The Lemma 2.5, giving the upper bounds for delay orders, and Lemma 2.10 were first proved by Nijmeijer [Nij87], but the proofs described here are the discrete-time analogues of the proofs given in [Isi89].

The definition of forward time-shift (FTS) right invertibility is actually a discrete-time analogue of the definition due to Respondek (see Definition 4.2 in [Res90]). The special case of FTS-right invertibility, the notion of $(d_1, \ldots, d_p)$-FTS right invertibility has been introduced in [Kot90]; necessary and sufficient conditions for $(d_1, \ldots, d_p)$-FTS right invertibility are taken from [Kot90]. Note that in the case of continuous-time systems, the equivalent notion of $(d_1, \ldots, d_p)$-FTS right invertibility is the notion of system with vector relative degree [Isi89]. The concepts of reduced-order and „state-free" (left) inverse system for discrete-time nonlinear systems are developed in [MNC87]. The usefulness of the Lie series in the solution of the system of nonlinear equations (which is required for construction of the right inverse system) was pointed out by Monaco, Normand-Cyrot and Isola [MNCI89]. The earlier results on discrete-time nonlinear system inversion, dealing with the very special subclasses of systems (2.1) may be found in [MNC83a], [Kot83], [Kot85], [Kot86a–c].

In our definition of right invertibility $x(0)$ is fixed. If we do allow the initial state $x(0)$ in (2.1) to vary, we obtain a slightly different notion of right invertibility, because sometimes we can satisfy the conditions (2.8) by a proper choice of $x(0) = x_0$. For continuous-time nonlinear systems the role played by the initial state in system inversion has been studied in [RN88], [Res90].

Note that the Examples 2.11 and 2.12 are taken from [BMS80] and [Pas89] respectively.

More notes and references on discrete-time system inversion can be found in Chapter 5.

We would like to end up with some remarks concerning the terminology around the problem of right invertibility which seems not to be quite standard. Besides the notion of right invertibility other notions such as functional reproducibility [BM65], [AGW86], [Nij89], functional controllability [SM69] and path controllability [Woh85], [AGW86], [Nij89] are used to denote the same property of the system. Throughout this book we use the notion of right invertibility. Note that the concept of output controllability [KS64] is different from the concept of right invertibility. Output controllability, unlike right invertibility, is a pointwise property, i.e. it is concerned with the existence of an input sequence that drives the output to a specified point in output space at a specified time instant. The concept of right invertibility, on the other hand, is a functional property, concerned with the existence of an input sequence which produces a specified output sequence (function).

[AGW86] Albrech F., K.A.Grasse and N.Wax. Path controllability of linear input-output systems. *IEEE Trans. Autom. Control*, 1986, v. 31, 569–571.

[BMS80] Baheti R.S., R.R.Mohler, H.A.Spang III. Second order correlation method for bilinear system identification. *IEEE Trans. Autom. Control*, 1980, v. 25, 1141–1146.

[BM65] Brockett R.W. and M.D.Mesarovic. The reproducibility of multivariable systems. *J. Math. Anal. Appl.*, 1965, v. 11, 548–563.

[Gra88] Grasse K.A. Sufficient conditions for the functional reproducibility of time-varying, input-output systems. *SIAM J. Contr. and Optimiz.*, 1988, v. 26, 230–249.

[Isi89] Isidori A. *Nonlinear Control Systems.* Berlin, Springer-Verlag, 1989.

[Kot83] Kotta Ü. On the inverse of a special class of MIMO bilinear systems. *Proc. Estonian Acad. Sci. Math., Phys.*, 1983, v. 32, 323-326.

[Kot85] Kotta Ü. Invertibility of bilinear discrete-time systems. *Proc. of IFAC/IFORS Conf. on Control Science and Technology for Development.* Beijing, 1985.

[Kot86a] Kotta Ü. Inversion of discrete-time linear-analytic systems. *Proc. Estonian Academy of Sci. Phys. Math.*, 1986, v. 35, 425–431.

[Kot86b] Kotta Ü. On the inverse of discrete-time linear-analytic system. *Control-Theory and Advanced Technology*, 1986, v. 2, 619–625.

[Kot86c] Kotta Ü. Construction of inverse system for discrete time nonlinear systems (In Russian). *Proc. Acad. Sci. of USSR. Technical Cybernetics*, 1986, 159–162.

[Kot90] Kotta Ü. Right inverse of a discrete time non-linear system. *Int. J. Control*, 1990, v. 51, 1–9.

[KS64] Kreindler E. and P.E.Sarachik. On the concepts of controllability and observability of linear systems. *IEEE Trans. Autom. Control*, 1964, v. 9, 129–136.

[MNC83a] Monaco S. and D.Normand-Cyrot. Some remarks on the invertibility of nonlinear discrete-time systems. *Proc. American Control Conference*, 1983, 229–245.

[MNC83b] Monaco S. and D.Normand-Cyrot. The immersion under feedback of a multidimensional discrete-time non-linear system into a linear system. *Int. J. Control*, 1983, v. 38, 245–261.

[MNC87] Monaco S. and D.Normand-Cyrot. Minimum-phase nonlinear discrete-time systems and feedback stabilization. *Proc. 26th IEEE Conf. on Desision and Control*, Los Angeles, CA, 1987, 979–986.

[MNCI89] Monaco S., D.Normand-Cyrot and T.Isola. Nonlinear decoupling in discrete time. *Prepr. of 1st IFAC Symp. on Nonlinear Control Systems Design*, Italy, Capri, 1989, 48–55.

[Nij87] Nijmeijer H. Local (dynamic) input-output decoupling of discrete-time nonlinear systems. *IMA J. of Mathematical Control and Information*, 1987, v. 4, 237–250.

[Nij89] Nijmeijer H. On dynamic decoupling and dynamic path controllability in economic systems. *Journal of Economic Dynamics and Control*, 1989, v. 13, 21–39.

[Pas89] Passino K.M. Disturbance rejection in nonlinear systems: examples. *IEE Proc. D.*, 1989, v. 136, 317–323.

[Res90] Respondek W. Right and left invertibility of nonlinear control systems. In: *Nonlinear Controllability and Optimal Control*, (ed. H.J.Sussmann), 1990, 133–176.

[RN88] Respondek W. and H.Nijmeijer. On local right invertibility of nonlinear control systems. *Control-Theory and Advanced Technology*, 1988, v. 4, 325–348.

[Sin82] Singh S.N. Functional reproducibility of multivariable nonlinear systems. *IEEE Trans. Autom. Contr.*, 1982, v. 27, 270–272.

[SM69] Sain M.K. and J.L.Massey. Invertibility of linear time-invariant dynamical systems. *IEEE Trans. Autom. Contr.*, 1969, v. 14, 141–149.

[Woh85] Wohtlmann H.W. Target path controllability of linear time-varying dynamical systems. *IEEE Trans. Autom. Control*, 1985, v. 30, 84–87.

3. The Inversion Method and Applications

In this chapter we shall assume that the system (1.11) is $(d_1, \ldots, d_p)$-forward time-shift right invertible, and explain the basic ideas of the inversion method for this special class of systems. A more general case is left for Chapter 6. In Section 3.1 the inversion method will be introduced as a method to solve the model matching problem (MMP).

In the MMP one considers the problem of designing a compensator for a nonlinear discrete-time system under which the input-output map of the compensated system becomes the same as that of a prespecified model. In general, the model is also assumed to be nonlinear. The MMP is a typical design problem in the sense that it plays a role in various other problems like the disturbance decoupling problem, the input-output (I/O) linearization and decoupling problems.

There are two possibilities in specifying the model—either we can fix the equations of the model completely, or, we can require only some structural properties of the model. For example, we may require that the model is either linear or decoupled but otherwise arbitrary. The case of a completely fixed model is considered in Sections 3.2 and 3.3, the case of linear model in Section 3.4 and the case of decoupled model in Section 3.5.

It should be stressed that the main theorems in this chapter are established under the assumption that the decoupling matrix of the system has a constant rank around an equilibrium point (x^0, u^0) of the system. Note that the rank is in general state and control dependent. This assumption is formalized in the notion of regularity of the equilibrium point (x^0, u^0) with respect to $(d_1, \ldots, d_p)$-forward time-shift right invertibility.

3.1 An Inversion Method for $(d_1, \ldots, d_p)$-forward Time-shift Right Invertible System

Consider a discrete-time nonlinear system S, described by equations of the form

$$\begin{aligned} x(t+1) &= f(x(t), u(t)), \; x(0) = x_0, \\ y(t) &= h(x(t)), \end{aligned} \tag{3.1}$$

where as before $x \in X$, $u \in U$, $y \in Y$ denote respectively the state, the control and the output.

Let a discrete-time nonlinear model M be also given, described by equations of the form

$$\begin{aligned} x^M(t+1) &= f^M(x^M(t), u^M(t)),\ x^M(0) = x_0^M, \\ y^M(t) &= h^M(x^M(t)), \end{aligned} \tag{3.2}$$

where the states $x^M(t)$ belong to an open subset X^M of R^{n^M}, the inputs $u^M(t)$ belong to an open subset U^M of R^m and the outputs $y^M(t)$ belong to an open subset Y^M of R^p, all for $0 \le t \le t_F$. The mappings f^M and h^M are supposed to be smooth.

The compensator C used to control the system S is a discrete-time nonlinear system described by equations of the form

$$\begin{aligned} x^C(t+1) &= f^C(x^C(t), x(t), u^M(t)),\ x^C(0) = x_0^C, \\ u(t) &= h^C(x^C(t), x(t), u^M(t)) \end{aligned} \tag{3.3}$$

with the state $x^C(t) \in X^C$, an open subset of R^{n^C} and f^C and h^C are smooth mappings.

The composition of (3.1) and (3.3), initialized at (x_0, x_0^C), is denoted by $S \circ C$:

$$\begin{aligned} x(t+1) &= f(x(t), h^C(x^C(t), x(t), u^M(t))) \\ x^C(t+1) &= f^C(x^C(t), x(t), u^M(t)) \\ y^{S\circ C}(t) &= h(x(t)). \end{aligned} \tag{3.4}$$

Let the I/O maps of the original system and the model be denoted by $\Sigma_{x_0}^S$ and $\Sigma_{x_0^M}^M$ respectively, and the I/O map of the compensator be denoted by $\Sigma_{x_0^C}^C$. The MMP is defined as follows. Given the system S and the model M, find a causal compensator C and a map $\xi : X^M \to X^C$ such that the I/O map of the compensated system $S \circ C$ coincides with the I/O map of the model; or equivalently, that

$$\Sigma_{(x_0, \xi(x_0^M))}^{S\circ C} = \Sigma_{x_0^M}^M \tag{3.5}$$

holds.

Our purpose in this chapter is to develop a method for solving the MMP under the assumption that the system S is $(d_1, \ldots, d_p)$-FTS right invertible. It would be desirable to achieve (3.5) via suitable choice of feedback, but for considered class of systems we are not able to do this, unless there exists certain correspondence between the initial states of the original system and the model. The crucial fact is that the ith output of the $(d_1, \ldots, d_p)$-FTS right invertible system for time instants $0 \le t \le d_i - 1$ is completely determined by the initial state x_0 and this property cannot be changed under any feedback. The best one can achieve is the coincidence of the ith output of the closed-loop system and the model starting from $t = d_i$. So, replacing (3.5) by

$$\operatorname{diag}\{\delta^{d_1}, \ldots, \delta^{d_p}\} \circ \Sigma_{(x_0, \xi(x_0^M))}^{S\circ C} = \operatorname{diag}\{\delta^{d_1}, \ldots, \delta^{d_p}\} \circ \Sigma_{x_0^M}^M \tag{3.6}$$

seems to be the best adapted to our theory. For $(d_1, \ldots, d_p)$-FTS right invertible system S there exists a causal inverse $\bar{S}_R^{-1}$ so that

$$\mathrm{diag}\{\delta^{d_1},\ldots,\delta^{d_p}\}\circ\Sigma^S_{x_0}\circ\Sigma^{\bar{S}_R^{-1}}_{\bar{x}_0^R}\circ\mathrm{diag}\{\delta^{d_1},\ldots,\delta^{d_p}\}=\mathrm{diag}\{\delta^{d_1},\ldots,\delta^{d_p}\}\circ\mathcal{I}_p$$

holds. Comparing the above equality with (3.6) leads to the compensator C with the I/O map

$$\Sigma^C_{\xi(x_0^M)}=\Sigma^{\bar{S}_R^{-1}}_{\bar{x}_0^R}\circ\mathrm{diag}\{\delta^{d_1},\ldots,\delta^{d_p}\}\circ\Sigma^M_{x_0^M}\,. \tag{3.7}$$

Really, $\Sigma^C_{\xi(x_0^M)}$ defined by (3.7) satisfies the equality

$$\mathrm{diag}\{\delta^{d_1},\ldots,\delta^{d_p}\}\circ\Sigma^S_{x_0}\circ\Sigma^C_{\xi(x_0^M)}=\mathrm{diag}\{\delta^{d_1},\ldots,\delta^{d_p}\}\circ\Sigma^M_{x_0^M}$$

or equivalently, taking into account the equality (1.12), the equality (3.6).

So, to obtain the desired control law via the inverse system, $(d_1,\ldots,d_p)$-forward time-shift right inverse system $\bar{S}_R^{-1}$ of S must be first determined. Then a solution of the MMP is obtained as the output of the right inverse system[1]

$$\begin{aligned}\bar{x}^R(t+1)&=f(\bar{x}^R(t),\varphi(\bar{x}^R(t),u^R(t)))\,,\ \bar{x}^R(0)=x(0)\,,\\ u(t)&=\varphi(\bar{x}^R(t),u^R(t))\,,\end{aligned} \tag{3.8}$$

if we feed into equations (3.8) the appropriate shifts of the outputs of the model system

$$u^R(t)=[y_1^M(t+d_1),\ldots,y_p^M(t+d_p)]^T \tag{3.9}$$

and replace $y_i^M(t+d_i)$, $i=1,\ldots,p$ in (3.9) by a function

$$h_i^{M,d_i}(x^M(t),u^M(t),\ldots,u^M(t+d_i-1))\,.$$

The latter can be found by applying to the ith output equation of the model

$$y_i^M(t)=h_i^M(x^M(t))$$

d_i times the forward shift operator δ like we did in computing the delay orders in Section 2.2. Note that we obtain a causal solution to the MMP if and only if the functions h_i^{M,d_i}, $i=1,\ldots,p$, do not depend on the future values of the model input $u^M(t+1),\ldots,u^M(t+d_i-1)$, or equivalently, the delay orders of the model are equal or greater than the corresponding delay orders of the system. Of course, the development sketched above is not possible if the system is not $(d_1,\ldots,d_p)$-forward time-shift right invertible.

The control law so obtained is not a state feedback but an open-loop compensator. The behaviour of the compensator C will be defined completely by the states x^M and $\bar{x}^R$ of the model M and the inverse system $\bar{S}_R^{-1}$ respectively, as well as by the model input u^M; it is not affected explicitly by the states x of the system S. This open-loop compensator can be replaced by an equivalent state feedback compensator. The replacement is based on a very simple observation. Namely, the states $\bar{x}^R$ of the right inverse system (3.8) and the states x of the original system

$$x(t+1)=f(x(t),u(t))$$

[1] For notational ease we drop the $(m-p)$-dimensional free parameter $\lambda(t)$ in the description of the right inverse system.

under the feedback $u(t) = \varphi(x(t), u^R(t))$, i.e. under the feedback we use, coincide, provided $\bar{x}^R(0) = x(0)$. So, the open-loop compensator can be interpreted as the feedback control if we just replace $\bar{x}^R(t)$ by $x(t)$ in the equations of the compensator. Note that making this replacement, there is no more need for the dynamic part of the right inverse (3.8) in the equations of the compensator, and we drop it.

The system design procedure using the inversion method may be summarized in the following algorithm.

1) Given S, compute a $(d_1, \ldots, d_p)$-forward time-shift right inverse $\bar{S}_R^{-1}$.

2) Choose the model, which corresponds to the design objectives and satisfies the condition $d_i^M \geq d_i$, $i = 1, \ldots, p$.

3) Compute the outputs of the model $y_i^M(t + d_i)$, $i = 1, \ldots, p$ in terms of the model state $x^M(t)$ and model input $u^M(t)$ using the state equations of the model
$$y_i^M(t + d_i) = h_i^{M,d_i}(f^M(x^M(t), u^M(t))).$$

4) The output equation of the compensator C can be found as the output equation of the right inverse system $\bar{S}_R^{-1}$ under the control
$$u^R(t) = h_i^{M,d_i}(f^M(x^M(t), u^M(t))), \; i = 1, \ldots, p).$$

5) The dynamic part of the compensator is the dynamic part of the model with $x^C(0) = x^M(0)$.

The methodology, described above, is simple to implement. The other advantage of this approach is that various results on inverses can be used here. Furthermore, after $\bar{S}_R^{-1}$ has been found, solutions C for different M's can be easily calculated. Note that in many problems M is not completely but only partially fixed and one may be interested in finding compensators for different models.

This approach also indicates how to choose M so that a causal solution C to (3.6) does exist. From (3.9) it is clear that the delay orders of the model must be equal or greater than the corresponding delay orders of the system. Only in that case $u^R(t)$ in (3.9) does not depend on future values of the reference input $u^M(t)$. Let us remark that in cases when the model is not completely specified like in problems of I/O linearization and I/O decoupling, one can choose M which meets this requirement.

So, the inversion method gives us both the conditions of the existence of solution and a method to find the solution. In addition, it also shows which structural properties can be achieved in the closed-loop.

3.2 The Formal Definition and the Solution of the Model Matching Problem

Let us first give the formal definition of the local MMP for $(d_1, \ldots, d_p)$-forward time-shift right invertible system.

Recall that we assume to work in a neighbourhood of an equilibrium point of the system (3.1), that is around a point $(x^0, u^0) \in X \times U$ such that $f(x^0, u^0) = x^0$. We say that the equilibrium point (x^{M0}, u^{M0}) of the model M is corresponding to the equilibrium point (x^0, u^0) of the system S if $y^0 = y^{M0} = h^M(x^{M0})$.

Definition 3.1 Local model matching problem. *Given the $(d_1, \dots, d_p)$-FTS invertible system S defined by equations (3.1) around an equilibrium point (x^0, u^0), the model M defined by equations (3.2) around an equilibrium point (x^{M0}, u^{M0}) corresponding to (x^0, u^0) and a point $(x(0), x^M(0))$, find if possible, a compensator C defined by equations of the form (3.3) together with an initial state $x^C(0)$, an equilibrium point x^{C0}, neighbourhoods $V_1 = X^0 \times X^{C0} \times U^{M0}$ of (x^0, x^{C0}, u^{M0}) in $X \times X^C \times U^M$ and V_2 of u^0 in U, being the domain and the range of C respectively, as well a neighbourhood X^{M0} of x^{M0} and a map $\xi : X^{M0} \to X^{C0}$ with the property that*

$$\begin{aligned} y_i^{S \circ C}(t, x(0), \xi(x^M(0)), u^M(0), \dots, u^M(t-1)) = \\ = y_i^M(t, x^M(0), u^M(0), \dots, u^M(t-1)), d_i \le t \le t_F\,,\ i = 1, \dots, p \end{aligned}$$

for all $(x(0), x^M(0)) \in X^0 \times X^{M0}$, and for all $u^M(t)$ in the domain of C.

Note that we do not require the first terms in the output sequences of $S \circ C$ and M (which are completely defined by the corresponding initial states) to be the same.

The next theorem is established under the assumption that we work around a regular (with respect to $(d_1, \dots, d_p)$-FTS right invertibility) equilibrium point (x^0, u^0) of the system S. Recall that regularity in the above sense means that the decoupling matrix of the system has a constant rank around an equilibrium point.

Theorem 3.2 *Consider the system (3.1) around a regular equilibrium point (x^0, u^0) and the model (3.2) around an equilibrium point (x^{M0}, u^{M0}), corresponding to (x^0, u^0). Suppose that the system (3.1) is locally around (x^0, u^0) $(d_1, \dots, d_p)$-forward time-shift right invertible. Then the model matching problem is locally solvable if and only if the delay orders of the model (3.2) are equal or greater than those of the original system: $d_i^M \ge d_i$, $i = 1, \dots, p$.*

Proof. The proof relies on the inversion method and on the Theorem 2.9. Let us consider the model around the equilibrium point (x^{M0}, u^{M0}) which corresponds to the regular equilibrium point (x^0, u^0) of the system (3.1). Then from (2.11) it follows that if and only if we apply a compensator C given by the equation

$$u(t) = \varphi(x(t), y_1^M(t + d_1), \dots, y_p^M(t + d_p)) \tag{3.10}$$

to the system (3.1), then the outputs of the model and the compensated system coincide (starting from time instant $t = d_i$ for the ith output component), i.e.

$$y_i^{S \circ C}(t) = y_i^M(t)\,,\ d_i \le t \le t_F\,,\ i \in \{1, \dots, p\}$$

as long as $(x(t), y_1^M(t + d_1), \dots, y_p^M(t + d_p)) \in \bar{X}^0 \times \bar{Y}^0$ and $u(t) \in \bar{U}^0$.

The compensator (3.10) can be given in the form (3.3) if and only if the delay orders d_i^M of the model are equal or greater than the corresponding delay orders of the system (3.1), that is $d_i^M \geq d_i$, $i = 1, \ldots, p$. In that case, defining the functions $h_i^{M,d_i}(x^M)$ analogously to the functions $h_i^{d_i}(x)$ (see Section 2.3) we obtain

$$y_i^M(t + d_i) = h_i^{M,d_i}(f^M(x^M(t), u^M(t))), \ i = 1, \ldots, p, \tag{3.11}$$

and substituting (3.11) into (3.10) gives

$$u(t) = \varphi(x(t), h_i^{M,d_i}(f^M(x^M(t), u^M(t))), i = 1, \ldots, p) \triangleq \bar{\varphi}(x(t), x^M(t), u^M(t)) .$$

Note that the states of the model x^M in this control law can be computed from the dynamic equations of the model which thus determine the dynamic equations of the compensator.

So, the compensator C together with its initial condition $x^C(0)$, solving the MMP is the following

$$\begin{aligned} x^C(t+1) &= f^M(x^C(t), u^M(t)), \ x^C(0) = x_0^M , \\ u(t) &= \bar{\varphi}(x(t), x^C(t), u^M(t)), \end{aligned} \tag{3.12}$$

and since $x_0^C = x_0^M$, $\xi = \mathrm{Id}$ (identity map).

From (3.11) we can see that $(x(t), y_1^M(t+d_1), \ldots, y_p^M(t+d_p^M))$ belong to $\bar{X}^0 \times \bar{Y}^0$ as long as $(x(t), x^C(t), u^M(t))$ belong to some neighbourhood V_1 of (x^0, x^{C0}, u^{M0}). This completes the proof. ∎

Let us remark that proof of the Theorem 3.2, up to application of the Implicit Function Theorem, is constructive. If we know how to solve equation (2.10), the proof indicates how to find the equations of the compensator which solves the MMP.

3.3 The Strong Model Matching Problem—Formulation and Solution

In the strong MMP (SMMP), unlike the Definition 3.1, the equality $y_i^{S \circ C}(t) = y_i^M(t)$ is required also for $0 \leq t \leq d_i - 1$, i.e. for the first terms in the output sequences of $S \circ C$ and M. As the values of $y_i^{S \circ C}(j)$, $j = 0, 1, \ldots, d_i - 1$, $i = 1, \ldots, p$ are completely defined by initial state of the system (3.1), the necessary condition for solvability of the SMMP is

(A1) $H^M(x^M(0)) = H(x(0))$,

where

$$\begin{aligned} H(x) &= \left[h_1^1(x), \ldots, h_1^{d_i}(x), \ldots, h_p^1(x), \ldots, h_p^{d_p}(x)\right]^T , \\ H^M(x^M) &= \left[h_1^{M,1}(x^M), \ldots, h_1^{M,d_1}(x^M), \ldots, h_p^{M,1}(x^M), \ldots, h_p^{M,d_p}(x^M)\right]^T . \end{aligned}$$

are $(d_1 + \ldots + d_p)$-dimensional vector functions.

If the SMMP is solvable, then it is possible to reduce the order of the dynamics of the compensator (3.12) and in some special cases even replace the dynamic compensator by static state feedback. We are going to show this.

From the coincidence of outputs of the closed-loop system and the model

$$y_i^{SoC}(t+j) = y_i^M(t+j)\,,\ 0 \le t+j \le t_F\,,\ i = 1,\dots,p\,,$$

in particular for $j = 0,\dots,d_i - 1$, it follows that for each $0 \le t \le t_F - d$, where $d = \max_i d_i$

$$H(x(t)) = H^M(x^M(t))\,. \tag{3.13}$$

Let us denote $d_1 + \ldots + d_p$ by μ. If the rank of the $(\mu \times n^M)$-dimensional matrix $\partial H^M(x^M)/\partial x^M$ is equal to μ (which by Lemma 2.10 is true for example for $(d_1,\dots,d_p)$-forward time-shift invertible models) then equation (3.13) can be solved for μ components of the vector x^M, denoted by x_1^M:[2]

$$x_1^M = \psi(x, x_2^M)\,,$$

in terms of x and the remaining $n^M - \mu$ components of x^M, denoted by x_2^M.

Substituting x_1^M into the equation $u(t) = \bar{\varphi}(x(t), x_1^M(t), x_2^M(t), u^M(t))$ we have to take into account only the dynamics of x_2^M which allows us to reduce the order of the compensator:

$$\begin{aligned} x^C(t+1) &= f_2^M(\psi(x(t), x^C(t)), x^C(t), u^M(t))\,,\ x^C(0) = x_2^M(0)\,,\\ u(t) &= \bar{\varphi}(x(t), \psi(x(t), x^C(t)), x^C(t), u^M(t))\,. \end{aligned}$$

Furthermore, if the following assumptions hold

(A2) $d_i = d_i^M\,,\ i = 1,\dots,p$

(A3) the order of the model n^M satisfies $n^M = d_1 + \ldots + d_p \triangleq \mu$

(A4) the $(\mu \times n^M)$-dimensional matrix $\partial H^M(x^M)/\partial x^M$ has the full rank: $\operatorname{rank}[\partial H^M(x^M)/\partial x^M] = \mu$,

then the state $x^M(t)$ of the model can be expressed by the state of the original system $x(t)$:

$$x^M(t) = (H^M)^{-1} \circ H(x(t))\,.$$

Substituting $x^M(t)$ in the equation $u(t) = \bar{\varphi}(x(t), x^M(t), u^M(t))$ gives us the compensator in the form of the static state feedback:

$$u(t) = \bar{\varphi}(x(t), (H^M)^{-1} \circ H(x(t)), u^M(t)) \triangleq \alpha(x(t), u^M(t))\,.$$

Summarizing, we have the following theorem.

---

[2]The vector $x_1^M = [x_{i_1}^M, \dots, x_{i_\mu}^M]$, $i_j \in \{1,\dots,n^M\}$, is chosen so that $\operatorname{rank}[\partial H^M(x^M)/\partial x_1^M] = \mu$ holds.

Theorem 3.3 *Consider the system (3.1) around a regular equilibrium point* (x^0, u^0) *and the model (3.2) around an equilibrium point* (x^{M0}, u^{M0})*, corresponding to* (x^0, u^0)*. Suppose that the system (3.1) is locally around* (x^0, u^0) $(d_1, \ldots, d_p)$*-FTS right invertible and that for (3.1) and (3.2) the conditions (A1)–(A4) hold. Then the strong model matching problem is locally solvable via the static state feedback of the form*

$$u(t) = \alpha(x(t), u^M(t)).$$

3.4 The Formulation and the Solution of the Input-output Linearization Problem via Static State Feedback

In this section we are interested in the problem of compensating the nonlinearities of the system (3.1), via feedback control. There are essentially two types of approaches to this problem. The first one consists in trying to find a state feedback (either static or dynamic) such that the dynamics of the closed loop system are, after a change of coordinates in the state space, locally equivalent to a linear system. The second approach tries to find a state feedback such that the input-output map of the closed-loop system becomes linear. In this book we are interested in the second approach, to be more precise, in linearizing the input-dependent part of the input-output map of the closed-loop system.

The input-output map of the system (3.1) around an equilibrium point (x^0, u^0) takes the form of a Volterra series expansion

$$y(t) = w^0(t, x_0) + \sum_{j\geq 1} \sum_{i_1, i_2, \ldots, i_j = 1}^{m} \sum_{\tau_1=0}^{t-1} \sum_{\tau_2=0}^{\tau_1} \cdots$$
$$\cdots \sum_{\tau_j=0}^{\tau_{j-1}} w^j_{i_1 i_2 \ldots i_j}(t, \tau_1, \ldots, \tau_j, x_0) \left[u_{i_1}(\tau_1) - u^0_{i_1}\right] \ldots \left[u_{i_j}(\tau_j) - u^0_{i_j}\right], \quad (3.14)$$

where $w^j_{i_1 i_2 \ldots i_j}$ is the jth triangular Volterra kernel.

There are two kinds of terms in the Volterra series (3.14): those that depend on the initial state x_0 alone and those that depend both on the controls u and the initial state x_0. Note that the kernels $w^j_{i_1 \ldots i_j}$ depend on the initial state x_0. Unlike the linear case, the response is not simply the sum of the forced and unforced responses.

The system (3.1) is said to have a linear input-output map if its Volterra series expansion reduces to one of the form

$$y(t) = w^0(t, x_0) + \sum_{i=1}^{m} \sum_{\tau=0}^{t-1} w^1_i(t - \tau)[u_i(\tau) - u^0_i], \; 0 \leq t \leq t_F, \quad (3.15)$$

that is if the input-dependent part of this expansion is

(i) linear in input u and

(ii) independent on the initial sate.

Note that the kernels w_i^1 in (3.15) unlike those of (3.14) do not depend on x_0. So, the response (3.15) is simply a sum of the forced and unforced responses. The term $w^0(t, x_0)$ in (3.15) which represents the response to the reference input $u(t) = u^0$, $0 \leq t \leq t_F$, need not necessarily be the same as the one of a linear system. However, this term affects the response to different inputs around reference input in the same way.

If the system is such that (3.15) is not true, we may try to satisfy this property via feedback, that is to find a state feedback such that for every $0 \leq t \leq t_F$ the input-dependent part of the input-output map of the closed-loop system is linear.

In this chapter we are looking for a static state feedback C with a new m-dimensional control v, described by equations of the form

$$u(t) = \alpha(x(t), v(t)) \tag{3.16}$$

defined locally around (to be found) a point (x^0, v^0, u^0) such that $u^0 = \alpha(x^0, v^0)$.

We call the compensator C described by equation (3.16) regular, if the matrix $\partial\alpha(x, v)/\partial v$ is nonsingular around a point (x^0, v^0, u^0).

The closed-loop system (3.1), (3.16), initialized at x_0, that is the system

$$\begin{aligned} x(t+1) &= f(x(t), \alpha(x(t), v(t)))\,,\ x(0) = x_0\,, \\ y(t) &= h(x(t))\,, \end{aligned} \tag{3.17}$$

is denoted by $S \circ C$.

The closed-loop system (3.17) is said to have a linear input-output map if its Volterra series expansion reduces to one of the form

$$y(t) = w^0(t, x_0) + \sum_{i=1}^{m} \sum_{\tau=0}^{t-1} w_i^1(t-\tau)[v_i(\tau) - v_i^0]\,,\ 0 \leq t \leq t_F\,. \tag{3.18}$$

Definition 3.4 Local static input-output linearization problem. *Given the system (3.1) around a regular equilibrium point (x^0, u^0, y^0) find, if possible, a regular static state feedback C defined by equations of the form (3.16) together with a point (x^0, v^0, u^0) and neighbourhoods $\Lambda = X^0 \times V^0$ of (x^0, v^0) and U^0 of u^0, being the domain and the range of C, respectively, so that the outputs of the closed-loop system*

$$y^{S \circ C}(t, x_0, v(0), \ldots, v(t-1)) = w^0(t, x_0) + \sum_{i=1}^{m} \sum_{\tau=0}^{t-1} w_i^1(t-\tau)[v_i(\tau) - v_i^0]\,,\ 0 \leq t \leq t_F$$

for every $x_0 \in X^0, v(k) \in V^0$, $0 \leq k \leq t-1$.

The problem of input-output (I/O) linearization can be considered as a special case of the MMP. If in the MMP the equations of the model are completely specified, in the case of I/O linearization we require only that the model is linear

$$\begin{aligned} x^M(t+1) &= Ax^M(t) + Bv(t)\,,\ x^M(0) = x_0^M\,, \\ y_i^M(t) &= c_i x^M(t)\,,\ i = 1, \ldots, p\,. \end{aligned} \tag{3.19}$$

The degree of freedom in a choice of model order, its initial state and the elements of the matrices $A, B, C = [c_1^T, \ldots, c_p^T]^T$ are limited only by the possibility to solve the problem. By the results of the previous section it means that the inequalities

$$d_i^M \geq d_i \,, \; i = 1, \ldots, p. \tag{3.20}$$

must hold.

As we can always find a linear model which satisfies the constraints (3.20), the I/O linearization problem for $(d_1, \ldots, d_p)$-FTS right invertible systems is always solvable.

As we shall see in the following, freedom in the choice of the linear model has one important consequence. For $(d_1, \ldots, d_p)$-FTS right invertible systems, the solution of the I/O linearization problem can be found in the class of more simple compensators than compensators of the form (3.3), namely in the class of static state feedback control laws of the form (3.16). The compensator of the form (3.16) is of course, a special case of the compensator (3.3), when the dimension of dynamic part is equal to zero, that is the compensator has no dynamic part.

In order to find a solution in the class of static state feedback compensators, the model must satisfy the following properties (see the previous section):

(A1') The initial state x_0^M of the model satisfies the following equation

$$x_0^M = (H^M)^{-1} H(x_0) \,,$$

where x_0 is the initial state of the system and

$$H(x) = \begin{bmatrix} h_1^1(x) \\ \cdots \\ h_1^{d_1}(x) \\ \cdots \\ h_p^1(x) \\ \cdots \\ h_p^{d_p-1}(x) \end{bmatrix}.$$

(A2') The delay orders of the model are equal to the corresponding delay orders of the system: $d_i^M = d_i$, $i = 1, \ldots, p$, which in case of linear model in equivalent to the following condition

$$c_i B = c_i AB = \ldots = c_i A^{d_i-2} B = 0 \,,$$
$$c_i A^{d_i-1} B \neq 0 \,, \; i = 1, \ldots, p.$$

(A3) The dimension n^M of the model satisfies $n^M = d_1 + \ldots + d_p$.

(A4') The matrix $\partial H^M(x^M)/\partial x^M$ has full rank which in case of linear model is equivalent to

$$\operatorname{rank}\frac{\partial}{\partial x^M}H^M x^M = \operatorname{rank}H^M = \operatorname{rank}\begin{bmatrix} c_1 \\ c_1 A \\ \cdots \\ c_1 A^{d_1-1} \\ \cdots \\ c_p \\ \cdots \\ c_p A^{d_p-1} \end{bmatrix} = n^M .$$

The following theorem holds.

Theorem 3.5 *Suppose that the system (3.1) is locally $(d_1, \ldots, d_p)$-forward time-shift right invertible around a regular equilibrium point (x^0, u^0). Then the input-output linearization problem is locally solvable around (x^0, u^0) by static state feedback (3.16).*

Proof. The proof relies on Theorem 2.9 and on the fact that one can always choose a linear model that satisfies the conditions (A1)–(A4). From (2.11) it follows, that if we apply the compensator C of the form

$$u(t) = \varphi(x(t), y_1^M(t+d_1), \ldots, y_p^M(t+d_p)) \tag{3.21}$$

for system (3.1), then the outputs of the compensated system and the model coincide, that is

$$y_i^{SoC}(t+d_i) = y_i^M(t+d_i)\,,\ 0 \le t+d_i \le t_F\,,\ i = 1, \ldots, p \tag{3.22}$$

as long as $(x(t), y_1^M(t+d_1), \ldots, y_p^M(t+d_p)) \in \bar{X}^0 \times \bar{Y}^0$ and $u(t) \in \bar{U}^0$. Next we show that under the chosen model the compensator C can be expressed in the form (3.16).

Really, under the condition (A1) we have

$$y_i^M(t+d_i) = c_i A^{d_i} x^M(t) + c_i A^{d_i-1} B v(t)\,,\ i = 1, \ldots, p. \tag{3.23}$$

Furthermore, combining the condition (A4) with the equality (3.22), to be more precise, with

$$y_i^{SoC}(t+j) = y_i^M(t+j)\,,\ j = 0, \ldots, d_i - 1\,,\ i = 1, \ldots, p$$

we have that for all $0 \le t \le t_F$

$$H(x(t)) = H^M x^M(t). \tag{3.24}$$

By the conditions (A2)–(A3)

$$x^M(t) = (H^M)^{-1} H(x(t)). \tag{3.25}$$

The proof is completed by showing that substituting (3.25) into (3.23) and then (3.23) into the equation of the compensator (3.21) gives the compensator in the form (3.16).

∎

Remark. The proof of the theorem is, up to application of the Implicit Function Theorem, constructive. The solution of the input-output linearization problem is as follows

$$u(t) = \varphi(x(t), c_i A^{d_i}(H^M)^{-1}H(x(t)) + c_i A^{d_i-1}Bv(t)\,,\; i = 1,\ldots,p)\,. \tag{3.26}$$

where the function φ is obtained via the solution of the set of nonlinear equations.

Remark. The $(d_1,\ldots,d_p)$-FTS right invertibility is sufficient for linearization of the I/O map of the system via static state feedback but not necessary. The necessary and sufficient condtions for solvability of the I/O linearization problem will be presented in Section 6 after generalization of the notion of $(d_1,\ldots,d_p)$-FTS right invertibility.

3.5 The Formulation and the Solution of the Input-output Decoupling Problem via Static State Feedback

The system (3.1) is said to be input-output decoupled if—after a possible reordering of the outputs—the ith input u_i only influences the ith output component y_i and none of the other outputs $y_i, j \neq i$. In the case the system does not possess the above property one may try to satisfy this property via state feedback compensator. In this chapter we are looking for a static state feedback C, described by equations of the form (3.16).

Definition 3.6 Local static input-output decoupling problem. *Given the system (3.1) around a regular equilibrium point (x^0, u^0), find if possible, a regular static state feedback C, defined by equations of the form (3.16) together with a point (x^0, v^0, u^0) and neighbourhoods $O = X^0 \times V^0$ of (x^0, v^0) in $X \times V$ and U^0 of u^0 in U, being the domain and the range of C so that the closed-loop system $S \circ C$, described by (3.1), (3.16) is input-output decoupled on $O \times U^0$, for all $0 \leq t \leq t_F$. That is the first p components $v_1^M, \ldots, v_p^M$ of the new control v^M influence independently the p outputs $y_1, \ldots, y_p$ and all other components $v_{p+1} \ldots v_m$ affect none of the outputs.*

In the input-output (I/O) decoupling problem one is looking for the state feedback under which the I/O behaviour becomes the same as that of some decoupled system. This problem can be also considered as the special case of the model matching problem where the equations of the model are not completely specified, but we require only that the I/O map of the model is decoupled. The degree of freedom in the choice of model is limited only by the possibility to solve the problem. By the results of the Section 3.3 it means that the inequalities (3.20) must hold. As we can always find the decoupled model which satisfies the constraints (3.20), the I/O decoupling problem for $(d_1,\ldots,d_p)$-forward time-shift right invertible systems is always solvable via dynamic state feedback of the form (3.3).

Analogously to the case of I/O linearization problem it is easy to show that the solution of I/O decoupling problem for $(d_1,\dots,d_p)$-FTS invertible systems can be found in the class of static state feedback control laws of the form (3.16).

The following theorem holds.

Theorem 3.7 *The system (3.1) with $m = p$ is locally around a regular equilibrium point (x^0, u^0) I/O decouplable by regular static state feedback of the form (3.21), if and only if the system (3.1) is locally $(d_1,\dots,d_p)$-forward time-shift right invertible around (x^0, u^0).*

Proof. *Sufficiency.* Choose a decoupled model that satisfies the conditions (A1)–(A4). We can easily verify that the assumptions (A2)–(A4) hold if the decoupled model is described by the equations

$$\begin{aligned} x^M_{i,k}(t+1) &= x^M_{i,k+1}(t)\,,\ k = 1,\dots,d_i - 1\,,\\ x^M_{i,d_i}(t+1) &= \psi_i(x^M_{i1}(t),\dots,x^M_{i,d_i}(t), v_i(t))\,,\\ y^M_i(t) &= x^M_{i1}(t)\,,\ i = 1,\dots,p\,. \end{aligned} \tag{3.27}$$

Furthermore, if we assume that $x^M_0 = H(x_0)$, then assumption (A1) holds, because $H^M = I_{n^M}$. The rest of the proof is analogous to the proof of Theorem 3.5 and is left to the reader.

Necessity. Suppose that the I/O decoupling problem for system (3.1) is locally solvable in a neighbourhood of the regular equilibrium point (x^0, u^0). It means that there exists a smooth regular feedback

$$u(t) = \alpha(x(t), v(t)) \tag{3.28}$$

defined around the points (x^0, v^0) and u^0 (with v^0 as the solution of the equation $u^0 = \alpha(x^0, v^0)$) such that the feedback modified system

$$\begin{aligned} x(t+1) &= f(x(t), \alpha(x(t), v(t))) \triangleq \tilde f(x(t), v(t))\\ y(t) &= h(x(t)) \triangleq \tilde h(x(t)) \end{aligned} \tag{3.29}$$

is I/O decoupled. Then we know from the analysis of delay orders that, in particular, for all k and all $1 \le i \le p$

$$\frac{\partial}{\partial v_j}\tilde h^k_i(\tilde f(x,v)) = 0$$

whenever $j \ne i$. The above condition written for $k = d_i$ shows that the decoupling matrix $\tilde K(x,v)$ of the closed-loop system has a diagonal structure. But we have also that (see Lemma 3.8 below)

$$\tilde K(x,v) = K(x,u)_{|u=\alpha(x,v)} \frac{\partial \alpha(x,v)}{\partial v}\,.$$

Thus, since the matrix $\partial\alpha/\partial v$ is nonsingular by regularity of the feedback (3.28), and each row of $K(x,u)$ is nonzero by construction, each row of $\tilde K(x,v)$ is nonzero.

$\tilde{K}(x,v)$ being diagonal, this implies that the p rows of $\tilde{K}(x,v)$ are linearly independent and so are the p rows of $K(x,u)$. ∎

Lemma 3.8 *Consider the system (3.1) around an equilibrium point (x^0,u^0), and let $u=\alpha(x,v)$ be an arbitrary analytic state feedback defined around the points (x^0,v^0) and u^0. Then around (x^0,u^0,v^0) we have that*

$$\tilde{K}(x,v) = K(x,u)\big|_{u=\alpha(x,v)} \frac{\partial}{\partial v}\alpha(x,v) \tag{3.30}$$

where $\tilde{K}(x,v)$ is the decoupling matrix of the feedback modified system (3.29).

Proof. The functions $h_i^1(x),\dots,h_i^{d_i}(x)$, depending only upon x and not on the control, are not altered by applying state feedback. So, the systems (3.1) and (3.29) define the same functions h_i^k, $k=1,\dots,d_i$, $i=1,\dots,p$. The ith row of the decoupling matrix $K(x,u)$ is determined as

$$\frac{\partial}{\partial u}h_i^{d_i}(f(x,u)) = \frac{\partial}{\partial x}h_i^{d_i}(x)\big|_{x=f(x,u)}\frac{\partial}{\partial u}f(x,u)\,.$$

On the other hand, the ith row of $\tilde{K}(x,v)$ is determined as

$$\frac{\partial}{\partial v}h_i^{d_i}(f(x,\alpha(x,v))) = \frac{\partial}{\partial x}h_i^{d_i}(x)\big|_{x=f(x,\alpha(x,v))}\frac{\partial}{\partial u}f(x,u)\big|_{u=\alpha(x,v)}\frac{\partial}{\partial v}\alpha(x,v)$$

These two expressions yield (3.30). ∎

Remark. Unlike the I/O linearization problem, the $(d_1,\dots,d_p)$-forward time-shift right invertibility is also necessary for the solvability of the I/O decoupling problem, if we work around a regular equilibrium point.

Notes and References

The inversion method was first explicitly suggested in [Kot87a], see also [Kot88]. The MMP was solved for single-input single-output (SISO) systems in [Kot91], and for the subclass of linear analytic systems in [Kot87a], [Kot90]. The solution presented here, is taken from [Kot92]. The I/O linearization problem in the discrete-time was first studied by Monaco and Normand-Cyrot for linear-analytic [MNC83a] and for SISO [MNC83b] systems. The problem of approximate I/O linearization has been addressed in [Kot89]. The general case (Theorem 3.5) has never appeared in the literature. Note, however, that the problem is closely related to the I/O decoupling problem and the solution of the latter problem is known to provide a solution to the I/O linearization problem. The I/O decoupling problem in the discrete-time was first studied for the very restricted subclass of linear-analytic systems [MNC84], [Kot85]. The complete solution of the I/O decoupling problems, via static state feedback, in the general case is due to Nijmeijer [Nij87]. Additional results can be found in [MNC89]. A geometric approach to the block I/O decoupling problem has been given in [Gri86]. The notes and references given in this chapter concern the subclass of $(d_1, \ldots, d_p)$-FTS right invertible systems. Notes and references on the MMP, I/O linearization and decoupling problems, in the general case, can be found in Chapter 6.

[Gri86] Grizzle J.W. Local input-output decoupling of discrete-time nonlinear systems. *Int. J. Contr.*, 1986, v. 43, 1517–1530.

[Kot85] Kotta Ü. Decoupling of discrete-time nonlinear systems by state feedback. *Prepr. of AFCET Congress "Automatique 85"*. Toulouse, 1985.

[Kot87a] Kotta Ü. The matching of a prespecified linear input-output behaviour in a discrete time nonlinear system. *Prepr. of 7th Int. Conf. on Control Systems and Computer Sci.*, Bucharest, 1987.

[Kot87b] Kotta Ü. Model matching of nonlinear discrete time systems. *Prepr. of 1st Int. Conf. on Industrial and Appl. Mathematics.* Paris, 1987.

[Kot88] Kotta Ü. Discrete-time linear-analytic system linearization and decoupling via application of right inverse system. *Proc. Estonian Acad. Sci. Phys. Math.*, 1988, v. 37, 257–262.

[Kot89] Kotta Ü. Approximate input-output linearization of discrete-time nonlinear systems. *Proc. Estonian Acad. Sci. Phys. Math.*, 1989, v. 38, 460–462.

[Kot90] Kotta Ü. Model matching of linear-analytic discrete time systems via dynamic state feedback. *Proc. Estonian Acad. Sci. Phys. Math.*, 1990, v. 39, 236–246.

[Kot91] Kotta Ü. Model matching of nonlinear single-input single-output discrete-time systems: formal and local solutions. *Proc. Estonian Acad. Sci. Phys. Math.*, 1991, v.40, 89–98.

[Kot92] Kotta Ü. Model matching of nonlinear discrete-time systems via dynamic state feedback. *Proc. Estonian Acad. Sci. Phys. Math.*, 1992, v. 41, 109–117.

[Nij87] Nijmeijer H. Local (dynamic) input-output decoupling of discrete-time nonlinear systems. *IMA J. of Mathematical Control and Information*, 1987, v. 4, 237–250.

[MNC83a] Monaco S., and D.Normand-Cyrot. The immersion under feedback of a multidimensional discrete-time non-linear system into a linear system. *Int. J. Control*, 1983, v. 38, 245–261.

[MNC83b] Monaco S., and D.Normand-Cyrot. Formal power series and input-output linearization of nonlinear discrete-time systems. *Proc. 22nd IEEE Conf. on Decision and Control*, San Antonio, 1883, 665–670.

[MNC84] Monaco S., and D.Normand-Cyrot. Sur la commande non interactive des systèmes non linéaires en temps discrets. *Lect. Notes in Contr. and Inf. Systems*, 1984, v. 63, 364–377.

4. Systems with Input Disturbances

In the preceding two chapters, disturbances were not taken into consideration. We concentrated on developing the methods for constructing the right inverse system and solving the MMP for systems without disturbances. The aim of this chapter is to extend the results of Chapters 2 and 3 to the systems of the form

$$\begin{aligned} x(t+1) &= f(x(t), u(t), w(t)), \ x(0) = x_0, \\ y(t) &= h(x(t)), \end{aligned} \tag{4.1}$$

where as in (1.1) $x \in X, u \in U, w \in W, y \in Y$ denote respectively the state, the control, the disturbance and the output. Like in the previous chapters we are assumed to work in a neighbourhood of an equilibrium point of the system.

4.1 Structural Characterization of a System with Disturbances

In Section 2.3, certain structural parameters, the so-called delay orders d_i, for system without disturbances were introduced. The number d_i determines the inherent delay between the input and ith output. The systems of the form (4.1) have two types of inputs—the controls and the disturbances. Therefore, for the system of the form (4.1) two sets of integer indices will be defined: the delay orders d_i with respect to the control u and the delay orders c_i with respect to the disturbance w. These indices will be defined analogously to the definitions of Section 2.3.

With each component of the output y_i we can associate a delay order d_i (refered also in the literature as characteristic number or relative order) with respect to the control u in the following manner.

Given an arbitrary initial state $x(t) \in X$, arbitrary $u(t) \in U$, arbitrary $w(t) \in W$ we can compute for $i = 1, 2, \ldots, p$ the derivative

$$\frac{\partial}{\partial u(t)} h_i^1(x(t+1)) = \frac{\partial}{\partial u(t)} h_i^1(f(x(t), u(t), w(t))),$$

where by definition we set $h_i^1(x) = h_i(x)$. From the analyticity of the system (4.1) it follows that either the vector $\partial h_i^1(f(x(t), u(t), w(t)))/\partial u(t)$ is nonzero for all $(x(t), u(t), w(t))$ belonging to an open and dense subset O_i of $X \times U \times W$ or this vector vanishes for all $(x(t), u(t), w(t)) \in X \times U \times W$. In the first case we define $d_i = 1$ whereas in the latter case we continue by observing that the function $h_i^1(f(x(t),u(t),w(t)))$ does not depend on $u(t)$ and so we may write $h_i^1(f(x(t),u(t),w(t))) = h_i^2(x(t),w(t))$ for some analytic h_i^2 on $X \times W$.

Next we compute in an analogous fashion

$$\frac{\partial}{\partial u(t)} h_i^2(x(t+1),w(t+1)) = \frac{\partial}{\partial u(t)} h_i^2(f(x(t),u(t),w(t)),w(t+1)).$$

If this vector is nonzero on an open and dense subset O_i of $X \times U \times W \times W$, we set $d_i = 2$; otherwise we continue with the function

$$h_i^3(x(t),w(t),w(t+1)) = h_i^2(f(x(t),u(t),w(t)),w(t+1)).$$

In this way the number d_i—if it exists—determines the inherent delay between the controls and the ith output no matter what the disturbances are. Namely, the control $u(t)$ possibly affects the ith output only after d_i steps, that is at the time instant $t+d_i$. In the case none of the iterated functions

$$h_i^k(x(t),u(t),w(t),w(t+1),w(t+2),\ldots,w(t+k-1)),\ k \geq 1,$$

depend on $u(t)$, we define $d_i = \infty$. When $d_i = \infty$ the ith output evolves in time independently from the control sequence applied to the system (4.1).

Note that the condition

$$\frac{\partial}{\partial u(t)} h_i^{d_i}(f(x(t),u(t),w(t)),w(t+1),\ldots,w(t+d_i)) \neq 0$$

need not to hold, and consequently $u(t)$ need not affect the ith output at $t+d_i$ if $(x(t),u(t),w(t),w(t+1),\ldots,w(t+d_i)) \notin O_i$.

With each component of the output y_i we can associate a delay order c_i with respect to the disturbance w in a similar manner. We can compute

$$\frac{\partial}{\partial w(t)} h_i^1(f(x(t),u(t),w(t))).$$

If this vector is nonzero on an open and dense subset M_i of $X \times U \times W$, we set $c_i = 1$, otherwise we continue with the function $h_i^2(x(t),u(t)) = h_i^1(f(x(t),u(t), w(t)))$. The number c_i—if it exists—determines the inherent delay between the disturbances and the ith output no matter what the controls are. Namely, the disturbance $w(t)$ possibly affects the ith output only after c_i steps, that is at time instant $t+c_i$.

The slight change in the proof of Lemma 2.5 shows that the following holds.

Lemma 4.1 *The finite delay orders, both with respect to the control and the disturbance, satisfy the inequalities:* $d_i \leq n$, $c_i \leq n$, $i = 1,\ldots,p$.

So, the delay orders d_i and c_i provide a measure of how fast is the dynamic effect of the control and the disturbance respectively, on the ith output component y_i. Thus, by comparing d_i and c_i, one can determine which input, the control u or the disturbance w, has more faster effect on the ith output component y_i. Consequently, one should intuitively expect a different nature of the solution of control problems depending on the relation between delay orders. The following classification seems meaningful:

A. $c_i > d_i \ i = 1, \ldots, p$.
B. $c_i \geq d_i \ i = 1, \ldots, p$ and $c_i = d_i$ for some $i \in \{1, \ldots, p\}$.
C. $c_i < d_i$ for some $i \in \{1, \ldots, p\}$.

It is clear that for class A, the effect of the disturbance on the system output is less fast than the effect of the control

$$y_i(t + d_i) = h_i^{d_i}(f(x(t), u(t), w(t))) = a_i(x(t), u(t)) \, , \ i = 1, \ldots, p. \tag{4.2}$$

All the useful information on how the disturbance changes the system output can be found in the system states.

On the other hand, in the case B, both the disturbance and the manipulated input affect the ith output in the same way; to be more precise, provided $c_i = d_i$ we have

$$y_i(t + d_i) = h_i^{d_i}(f(x(t), u(t), w(t)) = a_i(x(t), u(t), w(t)) \, . \tag{4.3}$$

Finally, for the case C, the disturbance affects the ith output earlier than the control, to be more precise, provided $c_i > d_i$ we have

$$y_i(t + d_i) = h_i^{d_i}(f(x(t), u(t), w(t)), w(t+1), \ldots, w(t + d_i - c_i)) \, . \tag{4.4}$$

4.2 The Notion of $(d_1, \ldots, d_p)$-forward Time-shift Right Invertibility for Systems with Disturbances

In case of systems with the disturbances one can distinguish between two versions of invertibility, and consequently between two types of inverse systems. In the first version invertibility is defined with regard to both types of inputs, the controls and the disturbances, whereas the other version considers disturbances as system parameters and invertibility is defined with regard to the controls only. In this section we are only interested in the second version of invertibility.

Let us introduce the definition of $(d_1, \ldots, d_p)$-forward time-shift right invertibility with respect to the controls for systems with the disturbances.

Definition 4.2 *One says that the system (4.1) is locally in the neighbourhood $X^0 \times U^0 \times W^0$ of the equilibrium point (x^0, u^0, w^0) $(d_1, \ldots, d_p)$-forward time-shift right invertible with respect to the control, if for arbitrary sequence $\{y_{\mathrm{ref}}(t) \, , \ 0 \leq t \leq t_F\}$ from the set $\mathcal{Y}^0$, for arbitrary $x(0) \in X^0$ and for arbitrary disturbance sequence $\{w(t), \ 0 \leq t \leq t_F\}$ from $\mathcal{W}^0$ there exists the sequence of controls $\{u_{\mathrm{ref}}(t), \ 0 \leq t \leq t_F\}$ from the set $\mathcal{U}^0$ with the property that*

$$y_i(t, x(0), u_{\mathrm{ref}}(0), \ldots, u_{\mathrm{ref}}(t-1), w(0), \ldots, w(t-1)) = y_{\mathrm{ref},i}(t) \, ,$$
$$d_i \leq t \leq t_F \, , \ i = 1, \ldots, p.$$

Assuming that each delay order d_i is finite we introduce the so-called decoupling matrix

$$K(x(t),u(t),w(t),\dots,w(t+\gamma)) = \frac{\partial}{\partial u(t)}A(x(t),u(t),w(t),\dots,w(t+\gamma)) =$$
$$= \frac{\partial}{\partial u(t)}\begin{bmatrix} h_1^{d_1}(f(x(t),u(t),w(t)),w(t+1),\dots,w(t+d_1-c_1)) \\ \dots \\ h_p^{d_p}(f(x(t),u(t),w(t)),w(t+1),\dots,w(t+d_p-c_p)) \end{bmatrix}$$

where $\gamma := \max\{(d_1-c_1),\dots,(d_p-c_p)\}$. From the definition of the d_i's the rows of this matrix are nonzero functions on an open and dense subset O of $X \times U \times W \times \times \dots \times W$.

In the sequel we need a notion of regularity of the equilibrium point which will be defined below.

Definition 4.3 *We call an equilibrium point (x^0,u^0,w^0) of the system (4.1) regular with respect to the $(d_1,\dots,d_p)$-forward time-shift right invertibility, if the rank of the decoupling matrix $K(x,u,w,\dots,w)$ of the system (4.1) is constant in a neighbourhood of $(x^0,u^0,w^0,\dots,w^0)$.*

The following theorem will give necessary and sufficient condition for local $(d_1,\dots,d_p)$-forward time-shift right invertibility with respect to the control around the regular equilibrium point (x^0,u^0,w^0).

Theorem 4.4 *The system (4.1) is locally around a regular equilibrium point (x^0,u^0,w^0) $(d_1,\dots,d_p)$-forward time-shift right invertible with respect to the control if and only if rank $K(x^0,u^0,w^0,\dots,w^0) = p$.*

Proof. Sufficiency. From the definition of delay orders d_i with respect to the control it is clear that

$$[y_1(t+d_1),\dots,y_p(t+d_p)]^T = A(x(t),u(t),w(t),\dots,w(t+\gamma)). \tag{4.5}$$

If we can solve this equation for $u(t)$, $0 \le t \le t_F - \max d_i$, after replacing $y_i(t+d_i)$, $i=1,\dots,p$ by $y_{\mathrm{ref},i}(t+d_i) \in Y_i^0$, then the system (4.1) is locally right invertible.

Observe that the Jacobian matrix of the right hand side of (4.5) with respect to the control u equals to the decoupling matrix $K(x,u,w,\dots,w)$. By the assumption of the theorem the rank of the decoupling matrix $K(x,u,w,\dots,w)$ is equal to p at the point $(x^0,u^0,w^0,\dots,w^0)$. So, we may apply the Implicit Function Theorem in order to solve the system of equations (4.5) with respect to the control vector u. After a possible reordering of the control components we may assume that the Jacobian matrix of the right hand side of (4.5) with respect to $u^1 = (u_1,\dots,u_p)^T$ around the point $(x^0,u^0,w^0,\dots,w^0)$ has full row rank p. Moreover, by the definition of an equilibrium point we have

$$y^0 = A(x^0,u^0,w^0,\dots,w^0).$$

Therefore, equation (4.5) can be solved for $u^1(t)$ uniquely around the point (x^0,u^0,w^0,y^0) by applying the Implicit Function Theorem. Define $u^2=(u_{p+1},\ldots\ldots,u_m)^T$. The Implicit Function Theorem says that in some (possible small) neighbourhood $\bar{X}^0\times\bar{U}^0\times\bar{Y}^0\times\bar{W}^0\times\ldots\times\bar{W}^0$ of $(x^0,u^0,y^0,w^0,\ldots,w^0)$ there exists an analytic function φ of variables $x(t),y_1(t+d_1),\ldots,y_p(t+d_p),w(t),\ldots,w(t+\gamma)$, and $u^2(t)$ i.e.

$$u^1(t)=\varphi(x(t),y_1(t+d_1),\ldots,y_p(t+d_p),w(t),\ldots,w(t+\gamma),u^2(t)) \tag{4.6}$$

which is such that

$$\varphi(x^0,y^0,w^0,\ldots,w^0,u^{20})=u^{10}$$

and

$$\begin{aligned}&[y_1(t+d_1),\ldots,y_p(t+d_p)]^T\equiv\\&\equiv A(x(t),\varphi(x(t),y_1(t+d_1),\ldots,y_p(t+d_p),w(t),\ldots,w(t+\gamma),u^2(t)),\\&\quad u^2(t),w(t),\ldots,w(t+\gamma))\,.\end{aligned} \tag{4.7}$$

Notice that the identy (4.7) is lost if we leave the neighbourhoods $\bar{X}^0\times\bar{Y}^0\times\times\bar{W}^0\times\ldots\times\bar{W}^0\times\bar{U}^{20}$ or $\bar{U}^{10}$.

Necessity. We omit the proof, because it is analogous to the proof of Theorem 2.9.

■

Remark. In the case of systems with the disturbances we must distinguish between two questions:

1) the existence of a control that gives $y(t)=y_{\text{ref}}(t)$ and

2) the possibility to compute such a control. The latter, in general, (provided the answer to first question is yes) depends on the availability of future values of disturbances.

4.3 Construction of $(d_1,\ldots,d_p)$-forward Time-shift Right Inverse System with Respect to Control for Systems with Input Disturbances

If we consider the disturbances as parameters we can define for system S described by (4.1) the input-output map (at x_0 and at $\{w_t\}$) written $\Sigma^S_{[x_0,\{w_t\}]}$, to be the function from $\mathcal{U}$ to $\mathcal{Y}$

$$\Sigma^S_{[x_0,\{w_t\}]}:\mathcal{U}\to\mathcal{Y}$$

which assigns to each control sequence the output sequence assuming $x(0)=x_0$ and $w(t)=w_t\,,\ t\geq 0$.

In this section we shall assume that the system S described by equations (4.1) is locally around (x^0,u^0,w^0) right invertible with respect to the control, i.e. the rank condition of Theorem 4.4 holds. Under this assumption we derive the equations of right inverse system with respect to the control for system (4.1), that is the equations of the system S_R^{-1}, whose input-output map satisfies the equation

$$\mathrm{diag}\{\delta^{d_1},\ldots,\delta^{d_p}\}\circ \Sigma^{S}_{[x_0,\{w_t\}]}\circ \Sigma^{S_R^{-1}}_{[x_0^R,\{w_t\}]} = \mathrm{diag}\{\delta^{d_1},\ldots,\delta^{d_p}\}\mathcal{I}_p\,.$$

The derivation is analogous to the case of systems without disturbances (see Section 2.6). Apply to the output equation of the system (4.1) the forward shift operator until it becomes explicitly dependent on the control $u(t)$. Doing this we reach equations (4.5). To get the equations for a right inverse, we have to be able to solve the equations (4.1), (4.5) with respect to $u(t)$ and $x(t+1)$ in terms of $x(t)$, $y_1(t+d_1),\ldots,y_p(t+d_p)$, and $w(t),\ldots,w(t+\gamma)$. The solution of (4.5) is given by (4.6) with arbitrary $u^2(t)$. Equation (4.6) defines the required control of the given system (which yields the reference output) in terms of the state, the future values of the reference output and the disturbances. We can take this equation as the output equation of the right inverse system. To obtain the dynamic part of the inverse, we must substitute (4.6) into (4.1)

$$\begin{aligned} x(t+1) = f(x(t),\varphi(x(t),y_1(t+d_1),\ldots,y_p(t+d_p),w(t),\ldots,w(t+\gamma),u^2(t)),\\ u^2(t),w(t))\,. \end{aligned} \tag{4.8}$$

As in Section 2.6 we can easily obtain from equations (4.6) and (4.8) the causal $(d_1,\ldots,d_p)$-forward time-shift right inverse system

$$\begin{aligned} \bar{x}(t+1) &= f(\bar{x}(t),\varphi(\bar{x}(t),u^R(t),w(t),\ldots,w(t+\gamma),\lambda(t)),\lambda(t),w(t)) \\ y^R(t) &= \begin{bmatrix} \varphi(\bar{x}(t),u^R(t),w(t),\ldots,w(t+\gamma),\lambda(t)) \\ \lambda(t) \end{bmatrix}, \end{aligned} \tag{4.9}$$

i.e. the system $\bar{S}_R^{-1}$ whose input-output map satisfies the equation

$$\mathrm{diag}\{z^{d_1},\ldots,z^{d_p}\}\circ \Sigma^{S}_{[x_0,\{w_t\}]}\circ \Sigma^{\bar{S}_R^{-1}}_{[\bar{x}_0^R,\{w_t\}]}\circ \mathrm{diag}\{z^{d_1},\ldots,z^{d_p}\} = \mathrm{diag}\{z^{d_1},\ldots,z^{d_p}\}\circ \mathcal{I}_p\,.$$

So, the $(d_1,\ldots,d_p)$-FTS right inverse with respect to the control of the system (4.1), in general, depends on the future values of the disturbances. Therefore, although the system (4.1) is right invertible, to compute the actual control $u_{\mathrm{ref}}(t)$ which yields $y_{\mathrm{ref}}(t)$, future values of disturbances are needed.

4.4 The Solution of the Model Matching Problem in the Presence of Unmeasurable Disturbances

If we consider the case of unmeasurable disturbances, the compensator cannot depend on the disturbances. So, a suitable compensator we must look for is a compensator of the form described by equations (3.3). This leads us to the following definition.

Definition 4.5 The local model matching problem in the precence of unmeasurable disturbances. *Given the system S described by equations (4.1) around an equilibrium point (x^0, u^0, w^0), the model M defined by equations (3.2) around an equilibrium point (x^{M0}, u^{M0}) corresponding to (x^0, u^0, w^0) and a point $(x(0), x^M(0))$, find, if possible, a compensator C defined by equations of the form (3.3) together with an initial state $x^C(0)$ and an equilibrium point x^{C0}, neighbourhoods $V_1 = X^0 \times X^{C0} \times U^{M0}$ of (x^0, x^{C0}, u^{M0}) in $X \times X^C \times U^M$, V_2 of u^0 in U, being the domain and the range of C respectively, as well a neighbourhood X^{M0} of x^{M0} and a map $\xi : X^{M0} \to X^{C0}$ with the property that*

$$y_i^{S\circ C}(t, x(0), \xi(x^M(0)), u^M(0), \dots, u^M(t-1), w(0), \dots, w(t-1)) =$$
$$= y_i^M(t, x^M(0), u^M(0), \dots, u^M(t-1)), \; d_i \le t \le t_F, \; i = 1, \dots, p$$

for all $(x(0), x^M(0)) \in X^0 \times X^{M0}$, for all $u^M(t)$ in the range of C, and for all $w(t) \in W$.

Next we formulate our result on the MMP in the presence of unmeasurable disturbances for $(d_1, \dots, d_p)$-forward time-shift right invertible systems.

Theorem 4.6 *Consider the system (4.1) around a regular equilibrium point (x^0, u^0, w^0) and the model (3.2) around an equilibrium point (x^{M0}, u^{M0}), corresponding to (x^0, u^0, w^0). Suppose that the system (4.1) is locally around (x^0, u^0, w^0) $(d_1, \dots, d_p)$-forward time-shift right invertible with respect to the control. Then the model matching problem in the presence of unmeasurable disturbances is locally solvable if the following two conditions hold.*

A. The delay orders with respect to the control of the model (3.2) are equal or greater than those of the original system (4.1): $d_i^M \ge d_i$, $i = 1, \dots, p$.

B. The delay orders of the system (4.1) with respect to the disturbances are greater than the corresponding delay orders with respect to the controls: $c_i > d_i$, $i = 1, \dots, p$.

Proof. The proof relies on the inversion method and on the Theorem 4.4. Let us consider the model (3.2) around an equilibrium point (x^{M0}, u^{M0}) which corresponds to the equilibrium pint (x^0, u^0, w^0) of the system (4.1). Then from (4.6) and (4.7) it follows that if we apply a compensator C given by the equation[1]

$$u(t) = \varphi(x(t), y_1^M(t + d_1), \dots, y_p^M(t + d_p), w(t), \dots, w(t + \gamma)) \tag{4.10}$$

to the system (4.1) then the outputs of the compensated system and the model coincide, i.e.

$$y_i^{S\circ C}(t + d_i) = y_i^M(t + d_i), \; 0 \le t \le t_F - d_i, \; i = 1, \dots, p$$

[1]For notational ease we drop the $(m-p)$-dimensional free parameter $u^2(t)$ in (4.6).

as long as $(x(t), y_1^M(t+d_1), \ldots, y_p^M(t+d_p), w(t), \ldots, w(t+\gamma)) \in \bar{X}^0 \times \bar{Y}^0 \times \bar{W}^0 \times \ldots \times \bar{W}^0$ and $u(t) \in \bar{U}^0$.

The compensator (4.10) can be given in the form (3.3) if the conditions A and B hold. If condition A holds, then

$$y_i^M(t+d_i) = h_i^{M,d_i}(f^M(x^M(t), u^M(t)) .$$

Otherwise $y_i^M(t+d_i)$ will depend on $u^M(t+1)$. If condition B holds, the compensator (4.10) does not depend on disturbances since then it reduces to

$$u(t) = \varphi(x(t), y_1^M(t+d_1), \ldots, y_p^M(t+d_p)) . \tag{4.11}$$

So, the compensator C and the mapping ξ, solving the MMP in the presence of unmeasurable disturbances are the following

$$\begin{aligned} x^C(t+1) &= f^M(x^C(t), u^M(t)), x^C(0) = x_0^M , \\ u(t) &= \varphi(x(t), h_i^{M,d_i}(f^M(x^C(t), u^M(t)),\ i = 1, \ldots, p) \triangleq \\ &\triangleq \bar{\varphi}(x(t), x^C(t), u^M(t)) , \end{aligned} \tag{4.12}$$

$\xi = \mathrm{Id}$ (identity map).

From (4.12) we can see that $(x(t),\ y_1^M(t+d_1), \ldots, y_p^M(t+d_p))$ belongs to $\bar{X}^0 \times \bar{Y}^0$ as long as $(x(t), x^C(t), u^M(t))$ belongs to some neighbourhood V_1 of (x^0, x^{C0}, u^{M0}). Theorem has been proved. ∎

Remark. Condition B is not necessary for solvability of the MMP in the presence of unmeasurable disturbances. We shall illustrate this remark by means of an example. Consider the nonlinear system

$$\begin{aligned} x_1(t+1) &= x_1(t)u_1(t) + w_1(t)u_1(t) \\ x_2(t+1) &= x_3(t)u_2(t) \\ x_3(t+1) &= x_3(t)w_2(t) \\ y_1(t) &= x_1(t),\ y_2(t) = x_2(t) \end{aligned} \tag{4.13}$$

and the model

$$\begin{aligned} x_1^M(t+1) &= 0 \\ x_2^M(t+1) &= u_1^M(t) \\ y_1^M(t) &= x_1^M(t),\ y_2^M(t) = x_2^M(t) . \end{aligned} \tag{4.14}$$

For system (4.13) $c_1 = d_1 = 1$. Nevertheless there exists a state-feedback

$$\begin{aligned} u_1(t) &= 0 \\ u_2(t) &= u_1^M(t)/x_3(t) \end{aligned} \tag{4.15}$$

which gives the solution to the MMP problem. The reason lies in the following. If $c_1 = d_1$, equation (4.5) that must be solved in order to find the feedback (4.12), depends on the disturbances. Consequently, in general, also the feedback depends on the disturbances. This is not the case for model (4.14) where we can find a solution for the system of equations

$$\begin{aligned} 0 &= x_1(t)u_1(t) + w_1(t)u_1(t) \\ u_1^M(t) &= x_3(t)u_2(t) \end{aligned} \tag{4.16}$$

which does not depend on $w(t)$. In order to avoid situations described in the above example, we should work around a stronger regularity assumption of an equilibrium point (x^0, u^0, w^0) than the one given by Definition 4.2. Namely, we should require that also the rank of the matrix

$$\frac{\partial}{\partial(w(t), \ldots, w(t+\gamma))} A(x(t), u(t), w(t), \ldots, w(t+\gamma))$$

is constant around an equilibrium point. Under this stronger regularity assumption the conditions A and B of Theorem 4.6 are also necessary.

In the strong MMP (SMMP) unlike the Definition 4.5, the equality $y_i^{SoC}(t) = y_i^M(t)$ is required also for $0 \le t \le d_i - 1$.

Under the conditions of Theorem 4.6 the SMMP is solvable if also the condition (A1) holds. If the SMMP is solvable, then it is possible to reduce the order of the dynamics of the compensator (4.12) and in some special cases even replace the dynamic compensator by a static state feedback. The latter is possible if the conditions (A1)–(A4) from Section 3.4 hold. In that case the control is defined by the equation

$$u(t) = \bar{\varphi}(x(t), (H^M)^{-1} \circ H(x(t)), u^M(t)) \triangleq \varphi^*(x(t), u^M(t)).$$

So, we have found the control laws that lead to elimination of the effect of unmeasurable disturbances on the output of the compensated system and at the same time provide the disturbance-free closed-loop system with desirable I/O map.

4.5 The Solution of the Disturbance Decoupling Problem (DDP)

The system (4.1) is said to be disturbance decoupled if its output y is not affected by the disturbance w. In the case the system does not possess the above property, one may try to satisfy this property via state feedback. In this chapter we are looking for a static state feedback C with a new m-dimensional control v, described either by equations of the form

$$u(t) = \alpha(x(t), v(t), w(t)) \tag{4.17}$$

or

$$u(t) = \alpha(x(t), v(t)) \tag{4.18}$$

depending on the fact if the measurements of the disturbances are available or not. If the measurements of the disturbances are available and incorporated in the controller, they allow improvement in the control quality or allow to solve the control problems under more relaxed assumptions. Note that (4.17) and (4.18) are defined locally around the points (x^0, v^0, w^0, u^0) and (x^0, v^0, u^0) respectively. We call the feedback (4.17) regular, if the matrix $\partial\alpha(\cdot)/\partial v$ is nonsingular around (x^0, v^0, w^0, u^0). The regularity of (4.18) is defined analogously.

In this section we solve the disturbance decoupling problem (DDP) under the additional assumption that the system is $(d_1, \ldots, d_p)$-FTS right invertible with respect to the control. In Chapter 7 we solve the disturbance decoupling problem without any assumption on the system to be controlled.

Necessary and sufficient conditions in terms of delay orders for local solvability of the DDP are proposed separately for the cases of unmeasurable and measurable disturbances with more relaxed conditions for the latter case.

Definition 4.7 The local disturbance decoupling problem in case of unmeasurable disturbances (DDPud). *Given the system S described by equations (4.1) around an equilibrium point (x^0, u^0, w^0), find, if possible, a regular static state feedback C defined by equations of the form (4.18) together with a point (x^0, v^0, u^0) and neighbourhoods $X^0 \times V^0$ of (x^0, v^0) in $X \times V$, U^0 of u^0 in U being the domain and the range of C respectively such that for all $0 \leq t \leq t_F$ the outputs $y(t)$ of the closed-loop system $S \circ C$ are independent of $w(t)$ for every $x_0 \in X^0$, all $v(t) \in V^0$, and all $w(t) \in W$.*

Next we shall present the solution to the DDPud.

Theorem 4.8 *Consider the system (4.1) around a regular equilibrium point (x^0, u^0, w^0). Suppose that the system (4.1) is locally $(d_1, \ldots, d_p)$-forward time-shift right invertible with respect to the control around (x^0, u^0, w^0). Then the DDP in case of unmeasurable disturbances is locally solvable if and only if the delay orders with respect to the disturbance are greater than the corresponding delay orders with respect to the control, that is, iff $c_i > d_i$, $i = 1, \ldots, p$.*

Proof. Sufficiency. The proof relies on Theorem 4.6 and on the fact that one can always find the model of the form (3.2)

$$\begin{aligned} x^M(t+1) &= f^M(x^M(t), v(t)),\ x^M(0) = x_0^M\,, \\ y^M(t) &= h^M(x^M(t)) \end{aligned}$$

such that the conditions (A1)–(A4) from the Section 3.4 hold, i.e.

(A1) $H(x(0)) = H^M(x^M(0))$,

(A2) $d_i^M = d_i$, $i = 1, \ldots, p$

(A3) the order of the model $n^M = d_1 + \ldots + d_p$

(A4) the matrix $\partial H^M(x^M)/\partial x^M$ has the full rank.

The following model, for example, satisfies those above mentioned conditions

$$
\begin{aligned}
x_{i,k}^M(t+1) &= x_{i,k+1}^M(t),\ k=1,\dots,d_i-1,\ i=1,\dots,p\\
x_{i,d_i}^M(t+1) &= v_i(t)\\
y_i^M(t) &= x_{i1}^M(t),\\
x^M(0) &\triangleq \left[x_{11}^M(0),\dots,x_{1d_1}^M(0),\dots,x_{p1}^M(0),\dots,x_{pd_p}^M(0)\right]^T = H(x(0)).
\end{aligned}
$$

Then by the proof of Theorem 4.6 the compensator

$$u(t) = \varphi(x(t), v_1(t),\dots,v_p(t)), \tag{4.19}$$

which has been obtained as the solution of the system of equations

$$
\begin{aligned}
v_1(t) &= h_1^{d_1}(f(x(t),u(t))),\\
&\dots\\
v_p(t) &= h_p^{d_p}(f(x(t),u(t))),
\end{aligned}
$$

implies

$$
\begin{aligned}
y_i(t+d_i) &= v_i(t),\ i=1,\dots,p\\
y_i(t+j) &= h_i^{j+1}(x(t)),\ i=1,\dots,p,\ j=0,\dots,d_i-1.
\end{aligned}
$$

Necessity. Now let us assume that there exists a regular compensator C of the form (4.18) for S that locally around a regular equilibrium point (x^0,u^0,w^0) solves the disturbance decoupling problem. Consider, for arbitrary i and for $1\le k\le d_i-1$

$$y_i(t+k) = h_i^k(f(x(t),u^0,w(t)),w(t),\dots,w(t+k-1)). \tag{4.20}$$

Since (4.18) solves the DDP for S, the w-dependence of (4.20) should disappear if we substitute (4.18) into (4.20). Since (4.20) does not depend on the control explicitly by the definition of d_i, it cannot depend on w either and therefore

$$y_i(t+d_i-1) = h_i^{d_i}(x(t)).$$

So, we have proved that c_i must be equal to or greater than d_i, i.e. $c_i\ge d_i$, $i=1,\dots,p$. Now consider

$$y_i(t+d_i) = h_i^{d_i}(f(x(t),u(t),w(t))),\ i=1,\dots,p. \tag{4.21}$$

If we plug the control $u(t)=\alpha(x(t),v(t))$ in (4.21), the equation does not depend on w any more, since $u(t)=\alpha(x(t),v(t))$ solves the DDP for S. This implies that either

$$\frac{\partial}{\partial w(t)} h_i^{d_i}(f(x(t),u(t),w(t))) = 0, \tag{4.22}$$

holds or that there exists the feedback $u=\alpha(x,v)$ such that

$$\frac{\partial h_i^{d_i}}{\partial x}\frac{\partial f}{\partial w}(x,\alpha(x,v),w) = 0.$$

The latter equality, of course, will imply the nonregularity of the feedback. Consequently, (4.22), or equivalently, $c_i > d_i$ holds. ∎

Remark. The condition $c_i > d_i$ is not any more necessary if we allow for a nonregular feedback. The reason is that even if $c_i = d_i$, the independence of $h_i^{d_i}(f(x(t), \alpha(x(t), v(t)), w(t))$ from w can sometimes be satisfied by the choice of such $u = \alpha(x, v)$ that

$$\frac{\partial h_i^{d_i}}{\partial x}\frac{\partial f}{\partial w}(x, \alpha(x, v), w) = 0 .$$

The latter equality means that u satisfies an equation $\xi(x, u, w) = 0$, not depending on v, which of course, will imply the nonregularity of the feedback. This can be illustrated by the following example:

$$\begin{aligned} x_1(t+1) &= x_1(t)u_1(t) + w_1(t)u_1(t) \\ x_2(t+1) &= x_3(t)u_2(t) \\ x_3(t+1) &= x_3(t)w_2(t) \\ y_1(t) &= x_1(t),\ y_2(t) = x_2(t) . \end{aligned}$$

Here

$$\begin{aligned} y_1(t+1) &= x_1(t)u_1(t) + w_1(t)u_1(t) \\ y_2(t+1) &= x_3(t)u_2(t) \end{aligned}$$

and so $d_1 = c_1 = 1$. Hence by Theorem 4.8 the DDP with unmeasurable disturbances is not solvable. Nevertheless, there exists a nonregular static state feedback

$$\begin{aligned} u_1(t) &= 0 \\ u_2(t) &= \frac{v_2(t)}{x_3(t)} \end{aligned}$$

which achieves decoupling between the disturbances and the outputs. The reason is that we can satisfy the requirement

$$\frac{\partial h_1}{\partial x}\frac{\partial f}{\partial w} = u_1 = 0$$

by the choice $u_1 = 0$.

The formulation of DDP with measurable disturbances is analogous to the case of Definition 4.7 except that now one allows the disturbance measurements $w(t)$ to be incorporated in the feedback (4.17).

Next we shall give a local solution to the DDP with measurable disturbances.

Theorem 4.9 *Consider the system (4.1) around a regular equilibrium point* (x^0, u^0, w^0). *Suppose that the system (4.1) is locally around* (x^0, u^0, w^0) $(d_1, \ldots, d_p)$*-forward time-shift right invertible with respect to the control. Then the DDP with measurable disturbances is locally solvable if and only if the delay orders with respect to the disturbance are equal to or greater than the corresponding delay orders with respect to the control, that is, iff* $c_i \geq d_i$, $i = 1, \ldots, p$.

Proof. The proof of Theorem 4.9 is quite analogous to the proof of Theorem 4.8 and will be omitted.

Note that under the feedback (4.17) the class of systems for which the DDP can be solved becomes larger if compared to the case with the feedback (4.18).

4.6 Classes of Disturbance Decoupling Control Laws

The specific function $u(t) = \varphi(x(t), v(t))$ defined by (4.19) in the proof of Theorem 4.8 is only one of the possible feedbacks that may be used to achieve a closed-loop system that is disturbance decoupled. The input-output map for the closed-loop system (4.1), (4.19) is determined by

$$y_i(t+d_i) = v_i(t)\,,\ i = 1,\dots,p \tag{4.23}$$
$$y_i(t+j) = h_i^{j+1}(x(t))\,,\ i = 1,\dots,p,\ j = 0,\dots,d_i - 1\,. \tag{4.24}$$

Since d_i is constant, the order of the difference equation (4.23) is invariant. A more general I/O map can be obtained by choosing

$$u(t) = \varphi(x(t), v_1^*(t),\dots,v_p^*(t)) \tag{4.25}$$

where

$$v_i^*(t) = \sum_{k=1}^{p}\left[\sum_{r=1}^{d_k} a_{ir}^k h_k^r(x(t)) + b_{ik}v_k(t)\right] \tag{4.26}$$

and a_{ir}^k, b_{ik} are arbitrary coefficients such that the $(p\times p)$-dimensional matrix with elements b_{ik} is nonsingular (in order to get the regular compensator) and where $v^*(t) = (v_1^*(t),\dots,v_p^*(t)) \in Y^0$ (which can be guaranteed by the appropriate choice of v^0). Then the input-output map of the ith disturbance decoupled subsystem is given by (4.24) and

$$y_i(t+d_i) = \sum_{k=1}^{p}\left[\sum_{r=1}^{d_k} a_{ir}^k y_k(t+r-1) + b_{ik}v_k(t)\right]. \tag{4.27}$$

Therefore, the output of (4.1), (4.25), (4.26) is invariant with respect to the disturbances and the control law (4.25), (4.26) is also a disturbance decoupling control law for the system (4.1). Moreover, the input-output map (4.27) is linear and by an appropriate selection of the constants a_{ir}^k, b_{ik} we can get a desired input-output characteristics.

We can choose the control law even more general than the one given by (4.25), (4.26):

$$u(t) = \varphi(x(t), \{\Phi_i(h_k^r(x(t)), k = 1,\dots,p, r = 1,\dots,d_k, v(t)), i = 1,\dots,p\}) \tag{4.28}$$

where $\Phi_i \in Y^0$ are such arbitrary analytic functions of their arguments, that the $p\times p$ matrix with elements $\partial\Phi_i(h_k^r(x), v)/\partial v_j$ is nonsingular for all $x(t) \in X^0$ and $v(t) \in V^0$. Now, by (4.6) and (4.7), the input-output map of the closed-loop system (4.1), (4.28) is given by (4.24) and

$$y_i(t+d_i) = \Phi_i(y_k(t+r-1), k = 1,\dots,p, r = 1,\dots,d_k\,, v(t)), i = 1,\dots,p\,. \tag{4.29}$$

We see that the output of this new feedback system is still disturbance decoupled. Thus the control law (4.28) is a disturbance decoupling control law.

Consider now the case of measurable disturbances. The proof of Theorem 4.9 will provide the following disturbance decoupling control law

$$u(t) = \varphi(x(t), v_1(t), \ldots, v_p(t), w(t)) \tag{4.30}$$

which has been obtained as the solution of the system of equations

$$\begin{aligned} v_1(t) &= h_1^{d_1}(f(x(t), u(t), w(t)) \\ &\cdots \\ v_p(t) &= h_p^{d_p}(f(x(t), u(t), w(t))\,. \end{aligned} \tag{4.31}$$

To get the larger class of decoupling control laws we can act similarly as in the case of unmeasurable disturbances. So, we can choose the class of control laws

$$u(t) = \varphi(x(t), \{\Phi_i(h_k^r(x(t)), k = 1, \ldots, p, r = 1, \ldots, d_k, v(t))\,, i = 1, \ldots, p\}, w(t)) \tag{4.32}$$

where $\Phi_i \in Y^0$ are such arbitrary analytic functions of their arguments that the $p \times p$-matrix with elements $\partial\Phi_i(h_k^r(x), v)/\partial v_j$ is of full row rank for all $x(t) \in X^0$ and $v(t) \in V^0$.

Notes and References

The definition of delay orders for systems with disturbances have been extended for the discrete-time nonlinear case in [Kot89] and [Kot91a]. In the above two papers the DDP, under the assumption of $(d_1, \ldots, d_p)$-FTS right invertibility of the system, has been solved for single input and for multi-input discrete-time nonlinear systems, respectively. The solvability conditions though obtained by different methods (inversion method versus variational approach) are in fact the discrete-time analogue of the conditions [SI84] for continuous-time systems. For the earlier results on the problem, see [Kot90].

Note that $(d_1, \ldots, d_p)$-FTS right invertibility assumption is not necessary for the solvability of the DDP via static state feedback. Grizzle [Gri85] studies the problem via the geometric method and establishes necessary and sufficient conditions for its local solvability. See also [Kot91b] which can be considered as the generalization of the results [Kot91a] via the inversion method to less restricted class of systems than $(d_1, \ldots d_p)$-FTS right invertible systems, namely to the class of systems, linearizable via static state feedback.

[Gri85] Grizzle J.W. Controlled invariance for discrete-time nonlinear systems with an application to the disturbance decoupling problem. *IEEE Trans. Autom. Contr.*, 1985, v. 30, 868–874.

[Kot89] Kotta Ü. The disturbance decoupling problem in nonlinear discrete-time systems. *Prepr. of IFAC Nonlinear Control System Design Symposium.* Italy, Capri, 1989, 59–63.

[Kot90] Kotta Ü. The disturbance decoupling problem for discrete-time nonlinear systems (In Russian). *Automatics and Telemechanics*, 1990, 20–27.

[Kot91a] Kotta Ü. The local disturbance decoupling problem for nonlinear discrete-time systems. *Prepr. of IMACS Symp. on Modelling and Control of Technological Systems.* Lille, 1991.

[Kot91b] Kotta Ü. The disturbance decoupling by static state feedback for nonlinear discrete-time systems. *Prepr. of Int. Workshop "Control System Synthesis: Theory and Applications"*. Novosibirsk, 1991, 37–42.

[SI84] Shima M. and Y.Isurugi. Variational system theory II. *Prepr. 9th IFAC World Congress.* Budapest, 1984, 63–68.

Part II

Control System Design for Partly or Completely Right Invertible Systems

In the previous part of the book (Chapters 3–4) it was assumed that the system is $(d_1, \ldots, d_p)$-forward time-shift right invertible. The inversion method, however, can be extended to the cases when this is not true. In Chapters 6 and 7 both the inversion method and the earlier results on control system design will be extended to wider class of systems—to forward time-shift right invertible systems, or even only to partly right invertible systems. The basic (main) tool in this generalization is the inversion algorithm.

The definition of forward time-shift (FTS) right invertible systems was given in Section 2.2. Chapter 5 gives necessary and sufficient conditions for a system to be FTS-right invertible and a systematic procedure for constructing right inverse system if the system is right invertible.

Let us note that while the book is concentrated on right invertible systems, application of the inversion method does not always require right invertibility of the system to be controlled. In some cases partial right invertibility is enough. The latter is true with respect to the problem of model matching, input-output linearization and disturbance decoupling. For the input-output decoupling problem however, the solvability is equivalent to system's right invertibility.

5. System Inversion. General Case

For each initial state $x_0 = x(0)$, the system S described by equations

$$\begin{aligned} x(t+1) &= f(x(t), u(t)) \\ y(t) &= h(x(t)) \end{aligned} \tag{5.1}$$

defines a mapping $\Sigma^S_{x_0} : \mathcal{U} \to \mathcal{Y}$ (see Section 2.1). In this chapter we are concerned with the problem of determining when the mapping $\Sigma^S_{x_0}$ is locally right invertible (surjective) and with the problem of finding a state representation of the inverse when it exists. Rather than first looking for conditions under which $\Sigma^S_{x_0}$ is right invertible for each x_0, we directly attack the problem of obtaining a representation for the inverse mapping when it exists.

As is clear from the first part of the book, for system S defined by (5.1) no possibility exist to reproduce at the output an arbitrary reference signal starting from time instant $t = 0$. We are able to reproduce the reference signals at the output with some time-shifts and the smallest possible value of the time-shift is d_i for ith output component. These smallest values can be realized if the system of equations

$$\begin{bmatrix} y_1(t+d_1) \\ \vdots \\ y_p(t+d_p) \end{bmatrix} = A(x(t), u(t)) \tag{5.2}$$

can be solved for $u(t)$ for arbitrary $[y_i(t+d_1), \ldots, y_p(t+d_p)]^T$. If this system of equations is not solvable, or equivalently, S is not $(d_1, \ldots, d_p)$-forward time-shift right invertible, then it is not possible to reproduce at each output an arbitrary reference signal starting from time instant d_i for the ith output component.

Note that we cannot solve the system of equations (5.2) for $u(t)$ in case of arbitrary left hand side if some components of the vector function $A(x, u)$, as functions of the control, depend functionally on the others, or equivalently, if

$$\operatorname{rank} \frac{\partial}{\partial u} A(x, u) < p.$$

The idea to generalize the notion of right invertibility is to represent the functionally dependent components via the independent ones and apply to the dependent equations the one-step forward time-shift operator and repeat the whole procedure (say α times) until we obtain a system of equations which can be solved for the control $u(t)$ in terms of $x(t)$ and $y(t+1), y(t+2), \ldots, y(t+\alpha)$ in case of arbitrary reference signal, or it will become clear that the latter is impossible. If it is possible to obtain a system of equations which can be solved for the control then we are able to reproduce at the ith output y_i an arbitrary reference signal starting from certain time instant $\gamma_i \geq d_i$ with $\gamma_i > d_i$ for some $j \in \{1, \ldots, p\}$.

5.1 The Inversion (Structure) Algorithm

In this section an inversion (structure) algorithm for constructing an inverse of a non-linear discrete-time system is presented. The word structure in another name for an inversion algorithm, that is the "structure algorithm", indicates that the algorithm is somehow related to the structure of the system. In system theoretic terms the inversion (structure) algorithm simply reorganizes the information contained in the system equations into a different form which is more efficient for system inversion. The main operations at each step of the inversion algorithm are the following.

1) Separation of the functionally independent and dependent output components. Note that since we are trying to obtain a system of equations that can be solved for u (and not for both x and u), we are interested in functional independence with respect to the control only and not with respect to all arguments.

2) Elimination the control from the functionally dependent output components expressing the latter via the independent components.

3) Application of the forward time-shift operator δ to the functionally dependent output components.

These operations when repeated will eventually allow to invert the original system provided the inverse exists. The action of the inversion algorithm is best understood by considering a few steps of the algorithm.

Now we shall present the inversion algorithm for system (5.1).

Denote $\hat{y}_0(t) = h(x(t))$ and $\rho_0 = 0$.

Step 1. Compute

$$y(t+1) = h(f(x(t), u(t))), \tag{5.3}$$

and define

$$\rho_1 = \text{rank} \frac{\partial}{\partial u} h(f(x,u))_{|x=x^0, u=u^0}.$$

Let us assume that rank $\partial h(f(x,u))/\partial u$ is const in some neighbourhood O_1 of (x^0, u^0). Reorder, if necessary, the system outputs $y_1, \ldots, y_p$ to ensure that the linearly independent rows of the matrix $\partial h(f(x,u))/\partial u$ are the first ρ_1 ones. Decompose $y(t+1)$ and $h(f(x,u))$ according to

$$y(t+1) = \begin{bmatrix} \tilde{y}_1(t+1) \\ \hat{y}_1(t+1) \end{bmatrix}, \quad h(f(x,u)) = \begin{bmatrix} \tilde{a}_1(x,u) \\ \hat{a}_1(x,u) \end{bmatrix},$$

where $\tilde{y}_1(t+1)$ and $\tilde{a}_1(x,u)$ consist of the first ρ_1 components of $y(t+1)$ and $h(f(x,u))$ respectively. The last $p - \rho_1$ rows of the matrix $\partial h(f(x,u))/\partial u$ are linearly dependent on the first ρ_1 rows, which means that $\hat{a}_1(x,u)$, as a function of u, depends functionally on $\tilde{a}_1(x,u)$ and can be expressed as a function of $\tilde{a}_1(x,u)$ and x, that is

$$\hat{a}_1(x(t),u(t)) = \psi_1(x(t),\tilde{a}_1(x(t),u(t))) = \psi_1(x(t),\tilde{y}_1(t+1)).$$

So, equation (5.3) can be locally around (x^0,u^0,y^0) split up into the following two equations

$$\begin{aligned}\tilde{y}_1(t+1) &= \tilde{a}_1(x(t),u(t))\\ \hat{y}_1(t+1) &= \psi_1(x(t),\tilde{y}_1(t+1))\end{aligned}$$

where the matrix $\partial\tilde{a}_1(x,u)/\partial u$ has full row rank ρ_1.

Denote $\tilde{a}_1(x,u)$ by $A_1(x,u)$.

Step $k+1 (k \geq 1)$. Suppose that in Steps 1 through $k, \tilde{y}_1(t+1), \tilde{y}_2(t+2), \ldots$ $\ldots, \tilde{y}_k(t+k), \hat{y}_k(t+k)$ have been defined so that

$$\begin{cases}\tilde{y}_1(t+1) = \tilde{a}_1(x(t),u(t))\\ \tilde{y}_2(t+2) = \tilde{a}_2(x(t),u(t),\tilde{y}_1(t+2))\\ \quad\ldots\\ \tilde{y}_k(t+k) = \tilde{a}_k(x(t),u(t),\{\tilde{y}_i(t+j),\ 1\leq i\leq k-1,\ i+1\leq j\leq k\})\end{cases} \tag{5.4}$$

$$\hat{y}_k(t+k) = \psi_k(x(t),\{\tilde{y}_i(t+j),\ 1\leq i\leq k,\ i\leq j\leq k\}). \tag{5.5}$$

Suppose also that the matrix $\partial A_k/\partial u = \partial[\tilde{a}_1^T\ldots\tilde{a}_k^T]^T/\partial u$ has full row rank equal to ρ_k in some neighbourhood O_k of (x^0,u^0). In the following we leave the first ρ_k equations unchanged and modify only the equations which do not depend explicitly on the control. Apply the forward time-shift operator δ upon (5.5)

$$\begin{aligned}\hat{y}_k(t+k+1) &= \psi_k(f(x(t),u(t)),\{\tilde{y}_i(t+j+1),\ 1\leq i\leq k,\ i\leq j\leq k\}) =\\ &\triangleq a_{k+1}(x(t),u(t),\{\tilde{y}_i(t+j),\ 1\leq i\leq k,\ i+1\leq j\leq k+1\})\end{aligned}$$

and define

$$\rho_{k+1} = \operatorname{rank}\frac{\partial}{\partial u}\begin{bmatrix}A_k(\cdot)\\ a_{k+1}(\cdot)\end{bmatrix}_{x=x^0,u=u^0,y=y^0=h(x^0)}.$$

Let us assume that rank $\partial[A_k^T,a_{k+1}^T]^T/\partial u$ is const in some neighbourhood O_{k+1} of (x^0,u^0). Reorder, if necessary, the components of $\hat{y}_k(t+k+1)$ and a_{k+1} so that the first ρ_{k+1} rows of the matrix $\partial[A_k^T,a_{k+1}^T]^T/\partial u$ are linearly independent. Note that this procedure does not change the rows of $\partial A_k/\partial u$, obtained at the previous steps. Decompose $\hat{y}_k(t+k+1)$ and a_{k+1} according to

$$\hat{y}_k(t+k+1) = \begin{bmatrix}\tilde{y}_{k+1}(t+k+1)\\ \hat{y}_{k+1}(t+k+1)\end{bmatrix}, \quad a_{k+1} = \begin{bmatrix}\tilde{a}_{k+1}\\ \hat{a}_{k+1}\end{bmatrix},$$

where $\tilde{y}_{k+1}(t+k+1)$ and $\tilde{a}_{k+1}$ consist of the first $\rho_{k+1}-\rho_k$ components of $\hat{y}_k(t+k+1)$ and a_{k+1} respectively. Since the last $p-\rho_{k+1}$ rows of the matrix $\partial[A_k^T,a_{k+1}^T]^T/\partial u$ are linearly dependent on the first ρ_{k+1} rows, we can write

$$\tilde{y}_1(t+1) = \tilde{a}_1(x(t), u(t))$$
$$\dots$$
$$\tilde{y}_{k+1}(t+k+1) = \tilde{a}_{k+1}(x(t), u(t), \{\tilde{y}_i(t+j),\ 1 \le i \le k,\ i+1 \le j \le k+1\})$$
$$\hat{y}_{k+1}(t+k+1) = \psi_{k+1}(x(t), \{\tilde{y}_i(t+j),\ 1 \le i \le k+1,\ i \le j \le k+1\}).$$

Denote

$$A_{k+1} = [A_k^T, \tilde{a}_{k+1}^T]^T.$$

End of step $k+1$.

Note that we can apply the inversion algorithm not necessarily in a unique way. There exist, in general, different reorderings (permutations) of output components $\hat{y}_k(t+k+1)$ at step $k+1, k \ge 0$, so that the first ρ_{k+1} rows of the matrix $\partial[A_k^T, a_{k+1}^T]^T/\partial u$ are linearly independent. Different permutations of output components, that is, different selections of $\tilde{y}_{k+1}(t+k+1)$ in each step $k+1, k \ge 0$, result in different functions $A_{k+1}(\cdot)$ and $\psi_{k+1}(\cdot)$.

In the inversion algorithm certain constant rank conditions have been imposed to ensure that the algorithm can be applied around a given equilibrium point. We shall summarize these conditions in the definition of regularity, associated with the inversion algorithm, of an equilibrium point.

Definition 5.1 *We call the equilibrium point (x^0, u^0) of the system (5.1) regular with respect to the inversion algorithm if for some specific application i of the inversion algorithm* rank $\partial[A_k^T, a_{k+1}^T]^T/\partial u$, $k \ge 1$, *is constant around (x^0, u^0). We call (x^0, u^0) strongly regular if the above assumptions hold for each application of the algorithm.*

The notion of regularity of an equilibrium point introduced in Definition 5.1 generalizes the one given by Definition 2.8.

5.2 Invertibility Indices, Rank and Tracking Order of the System

Through the result of application of the inversion algorithm apparently depends on the choice of admissible permutations made at each step of the algorithm, we have the following Lemma.

Lemma 5.2 *Around a strongly regular equilibrium point the integers $\rho_k^i, k \ge 1$, do not depend on the specific application i of the algorithm, that is $\rho_k^{i_1} = \rho_k^{i_2}, k \ge 1$, for each i_1 and i_2.*

Proof. We shall prove that the selected ordering of the output components at the first step of the algorithm does not influence the integer ρ_2, defined at the second step. The proof for the next steps may be handled in much the same way.

Let us consider two specific orderings of the components of the vector $y(t+1)$, where the second is obtained from the first by exchange the ρ_1-th and the

$(\rho_1 + 1)$th components (every other permutation can be obtained as a sequence of such elementary permutations and does not cause additional difficulties):

$$\tilde{y}_1^T = [y_1, \ldots, y_{\rho_1 - 1}, y_{\rho_1}] = [\tilde{\tilde{y}}_1, y_{\rho_1}],$$
$$\hat{y}_1^T = [y_{\rho_1+1}, y_{\rho_1+2}, \ldots, y_p] = [\hat{y}_{11}, \hat{\hat{y}}_1],$$

and

$$\tilde{y}_1^{*T} = [y_1, \ldots, y_{\rho_1 - 1}, y_{\rho_1+1}] = [\tilde{\tilde{y}}_1^*, y_{\rho_1+1}],$$
$$\hat{y}_1^{*T} = [y_{\rho_1}, y_{\rho_1+2}, \ldots, y_p] = [\hat{y}_{11}^*, \hat{\hat{y}}_1^*].$$

In correspondence with these notations we introduce the following ones

$$\tilde{a}_1^T = [a_1, \ldots, a_{\rho_1 - 1}, a_{\rho_1}] = [\tilde{\tilde{a}}_1, a_{\rho_1}],$$
$$\hat{a}_1^T = [a_{\rho_1+1}, a_{\rho_1+2}, \ldots, a_p] = [a_{\rho_1+1}, \hat{\hat{a}}_1],$$
$$\psi_1^T = [\psi_{11}, \bar{\psi}_1],$$

and

$$\tilde{a}_1^{*T} = [a_1, \ldots, a_{\rho_1 - 1}, a_{\rho_1+1}] = [\tilde{\tilde{a}}_1^*, a_{\rho_1+1}],$$
$$\hat{a}_1^{*T} = [a_{\rho_1}, a_{\rho_1+2}, \ldots, a_p] = [a_{\rho_1}, \hat{\hat{a}}_1^*],$$
$$\psi_1^{*T} = [\psi_{11}^*, \bar{\psi}_1^*].$$

Observe that

$$\tilde{\tilde{y}}_1 = \tilde{\tilde{y}}_1^*, \; \tilde{\tilde{a}}_1 = \tilde{\tilde{a}}_1^*. \tag{5.6}$$

The permutation is admissible, if

$$\rho_1 = \operatorname{rank}\left[\left(\frac{\partial \tilde{\tilde{a}}_1}{\partial u}\right)^T, \left(\frac{\partial \hat{a}_1}{\partial u}\right)^T\right]^T = \operatorname{rank}\frac{\partial \tilde{a}_1}{\partial u} = \operatorname{rank}\frac{\partial \tilde{a}_1^*}{\partial u},$$

which implies

$$\frac{\partial a_{\rho_1+1}}{\partial u} = [\alpha_1(x,u), \ldots, \alpha_{\rho_1-1}(x,u), \alpha_{\rho_1}(x,u)]\frac{\partial \tilde{a}_1}{\partial u} \triangleq [\bar{\alpha}(x,u), \alpha_{\rho_1}(x,u)]\frac{\partial \tilde{a}_1}{\partial u}, \tag{5.7}$$

where $\alpha_{\rho_1}(x,u) \neq 0$.

We must show that

$$\rho_2 = \operatorname{rank} B = \operatorname{rank} B^*,$$

where

$$B = \frac{\partial}{\partial u}\begin{bmatrix} \tilde{a}_1(x,u) \\ \psi_1(f(x,u), \tilde{y}_1) \end{bmatrix}, \quad B^* = \frac{\partial}{\partial u}\begin{bmatrix} \tilde{a}_1^*(x,u) \\ \psi_1^*(f(x,u), \tilde{y}_1^*) \end{bmatrix}.$$

By differentiating the identities

$$\hat{a}_1(x,u) = \psi_1(x, \tilde{a}_1(x,u)),$$
$$\hat{a}_1^*(x,u) = \psi_1^*(x, \tilde{a}_1^*(x,u))$$

with respect to x, we obtain

$$\frac{\partial \hat{a}_1}{\partial x} = \frac{\partial \psi_1}{\partial x} + \frac{\partial \psi_1}{\partial \tilde{y}_1}\frac{\partial \tilde{a}_1}{\partial x}, \tag{5.8}$$

$$\frac{\partial \hat{a}_1^*}{\partial x} = \frac{\partial \psi_1^*}{\partial x} + \frac{\partial \psi_1^*}{\partial \tilde{y}_1^*}\frac{\partial \tilde{a}_1^*}{\partial x}. \tag{5.9}$$

Taking into account (5.6)–(5.9) we obtain

$$B = \begin{bmatrix} \dfrac{\partial \tilde{\tilde{a}}_1}{\partial u} \\ \dfrac{\partial a_{\rho_1}}{\partial u} \\ \begin{pmatrix} \dfrac{\partial a_{\rho_1+1}}{\partial x} - \dfrac{\partial \psi_{11}}{\partial \tilde{\tilde{y}}_1}\dfrac{\partial \tilde{\tilde{a}}_1}{\partial x} - \dfrac{\partial \psi_{11}}{\partial y_{\rho_1}}\dfrac{\partial a_{\rho_1}}{\partial x} \\ \dfrac{\partial \hat{\bar{a}}_1}{\partial x} - \dfrac{\partial \bar{\psi}_1}{\partial \tilde{\tilde{y}}_1}\dfrac{\partial \tilde{\tilde{a}}_1}{\partial x} - \dfrac{\partial \bar{\psi}_1}{\partial y_{\rho_1}}\dfrac{\partial a_{\rho_1}}{\partial x} \end{pmatrix}\dfrac{\partial f}{\partial u} \end{bmatrix}, \tag{5.10}$$

$$B^* = \begin{bmatrix} \dfrac{\partial \tilde{\tilde{a}}_1}{\partial u} \\ \bar{\alpha}\dfrac{\partial \tilde{\tilde{a}}_1}{\partial u} + \alpha_{\rho_1}\dfrac{\partial a_{\rho_1}}{\partial u} \\ \begin{pmatrix} \dfrac{\partial a_{\rho_1}}{\partial x} - \dfrac{\partial \psi_{11}^*}{\partial \tilde{\tilde{y}}_1^*}\dfrac{\partial \tilde{\tilde{a}}_1}{\partial x} - \dfrac{\partial \psi_{11}^*}{\partial y_{\rho_1+1}}\dfrac{\partial a_{\rho_1+1}}{\partial x} \\ \dfrac{\partial \hat{\bar{a}}_1}{\partial x} - \dfrac{\partial \bar{\psi}_1^*}{\partial \tilde{\tilde{y}}_1^*}\dfrac{\partial \tilde{\tilde{a}}_1}{\partial x} - \dfrac{\partial \bar{\psi}_1^*}{\partial y_{\rho_1+1}}\dfrac{\partial a_{\rho_1+1}}{\partial x} \end{pmatrix}\dfrac{\partial f}{\partial u} \end{bmatrix}. \tag{5.11}$$

Let us express the elements of the matrix $\partial\psi_1^*/\partial\tilde{y}_1^*$ via the entries of the matrix $\partial\psi_1/\partial\tilde{y}_1$. For this, differentiate the identity

$$\hat{a}_1^*(x,u) = \psi_1^*(x, \tilde{a}_1^*(x,u))$$

with respect to u taking into account (5.6):

$$\frac{\partial a_{\rho_1}}{\partial u} = \frac{\partial \psi_{11}^*}{\partial \tilde{\tilde{y}}_1^*}\frac{\partial \tilde{\tilde{a}}_1}{\partial u} + \frac{\partial \psi_{11}^*}{\partial y_{\rho_1+1}}\frac{\partial a_{\rho_1+1}}{\partial u}, \tag{5.12}$$

$$\frac{\partial \hat{\bar{a}}_1}{\partial u} = \frac{\partial \bar{\psi}_1^*}{\partial \tilde{\tilde{y}}_1^*}\frac{\partial \tilde{\tilde{a}}_1}{\partial u} + \frac{\partial \bar{\psi}_1^*}{\partial y_{\rho_1+1}}\frac{\partial a_{\rho_1+1}}{\partial u}. \tag{5.13}$$

Analogously, from the identity

$$\hat{a}_1(x,u) = \psi_1(x, \tilde{a}_1(x,u))$$

we obtain

$$\frac{\partial a_{\rho_1+1}}{\partial u} = \frac{\partial \psi_{11}}{\partial \tilde{\tilde{y}}_1}\frac{\partial \tilde{\tilde{a}}_1}{\partial u} + \frac{\partial \psi_{11}}{\partial y_{\rho_1}}\frac{\partial a_{\rho_1}}{\partial u}, \tag{5.14}$$

$$\frac{\partial \hat{\bar{a}}_1}{\partial u} = \frac{\partial \bar{\psi}_1}{\partial \tilde{\tilde{y}}_1}\frac{\partial \tilde{\tilde{a}}_1}{\partial u} + \frac{\partial \bar{\psi}_1}{\partial y_{\rho_1}}\frac{\partial a_{\rho_1}}{\partial u}. \tag{5.15}$$

From (5.13) and (5.15), taking into account (5.7) we obtain

$$\frac{\partial \bar{\psi}_1^*}{\partial \tilde{\tilde{y}}_1^*}\frac{\partial \tilde{\tilde{a}}_1}{\partial u} = \left(\frac{\partial \bar{\psi}_1}{\partial \tilde{\tilde{y}}_1} - \frac{\partial \bar{\psi}_1}{\partial y_{\rho_1+1}}\bar{\alpha}\right)\frac{\partial \tilde{\tilde{a}}_1}{\partial u} + \left(\frac{\partial \bar{\psi}_1}{\partial y_{\rho_1}} - \frac{\partial \bar{\psi}_1}{\partial y_{\rho_1+1}}\alpha_{\rho_1}\right)\frac{\partial a_{\rho_1}}{\partial u}.$$

Let us multiply both sides of the last equality with the right inverse matrix $(\partial\tilde{a}_1/\partial u)^+$ of the full row rank matrix $\partial\tilde{a}_1/\partial u$. As

$$\frac{\partial\tilde{a}_1}{\partial u}\left(\frac{\partial\tilde{a}_1}{\partial u}\right)^+ = \left[I_{\rho_1-1} \,\vdots\, 0\right],$$

$$\frac{\partial a_{\rho_1}}{\partial u}\left(\frac{\partial\tilde{a}_1}{\partial u}\right)^+ = [0\ldots0\;1],$$

we obtain

$$\left[\frac{\partial\bar{\psi}_1^*}{\partial\tilde{\bar{y}}_1^*} \,\vdots\, 0\right] = \left[\frac{\partial\bar{\psi}_1}{\partial\tilde{\bar{y}}_1} - \frac{\partial\bar{\psi}_1}{\partial y_{\rho_1+1}}\bar{\alpha} \,\vdots\, \frac{\partial\bar{\psi}_1}{\partial y_{\rho_1}} - \frac{\partial\bar{\psi}_1}{\partial y_{\rho_1+1}}\alpha_{\rho_1}\right],$$

which implies, that

$$\frac{\partial\bar{\psi}_1^*}{\partial y_{\rho_1+1}} = \frac{\partial\bar{\psi}_1}{\partial y_{\rho_1}}\frac{1}{\alpha_{\rho_1}}, \tag{5.16}$$

$$\frac{\partial\bar{\psi}_1^*}{\partial\tilde{\bar{y}}_1^*} = \frac{\partial\bar{\psi}_1}{\partial\tilde{\bar{y}}_1} + \frac{\partial\bar{\psi}_1}{\partial y_{\rho_1}}\frac{\bar{\alpha}}{\alpha_{\rho_1}}. \tag{5.17}$$

Substituting (5.7) in (5.12) and (5.14), and multiplying both equalities by the matrix $(\partial\tilde{a}_1/\partial u)^+$, we obtain:

$$\left[\frac{\partial\psi_{11}^*}{\partial\tilde{\bar{y}}_1} \,\vdots\, 0\right] + \left[\frac{\partial\psi_{11}^*}{\partial y_{\rho_1+1}}\bar{\alpha} \,\vdots\, 0\right] + \left[0 \,\vdots\, \frac{\partial\psi_{11}^*}{\partial y_{\rho_1+1}}\alpha_{\rho_1}\right] = [0\ldots0\;1],$$

$$\left[\frac{\partial\psi_{11}}{\partial\tilde{\bar{y}}_1} \,\vdots\, 0\right] + \left[0 \,\vdots\, \frac{\partial\psi_{11}}{\partial y_{\rho_1}}\right] = [\bar{\alpha}\;0] + [0\;\alpha_{\rho_1}],$$

which implies that

$$\frac{\partial\psi_{11}^*}{\partial y_{\rho_1+1}} = \frac{1}{\alpha_{\rho_1}} \tag{5.18}$$

$$\frac{\partial\psi_{11}^*}{\partial\tilde{\bar{y}}_1} = -\frac{\bar{\alpha}}{\alpha_{\rho_1}}, \tag{5.19}$$

$$\frac{\partial\psi_{11}}{\partial\tilde{\bar{y}}_1} = \bar{\alpha}, \tag{5.20}$$

$$\frac{\partial\psi_{11}}{\partial y_{\rho_1}} = \alpha_{\rho_1}. \tag{5.21}$$

Taking into account (5.20) and (5.21) we rewrite (5.10)

$$B = \begin{bmatrix} \dfrac{\partial\tilde{\tilde{a}}_1}{\partial u} \\ \dfrac{\partial a_{\rho_1}}{\partial u} \\ \begin{pmatrix} \dfrac{\partial a_{\rho_1+1}}{\partial x} - \bar{\alpha}\dfrac{\partial\tilde{\tilde{a}}_1}{\partial x} - \alpha_{\rho_1}\dfrac{\partial a_{\rho_1}}{\partial x} \\ \dfrac{\partial\hat{\tilde{a}}_1}{\partial x} - \dfrac{\partial\bar{\psi}_1}{\partial\tilde{\bar{y}}_1}\dfrac{\partial\tilde{\tilde{a}}_1}{\partial x} - \dfrac{\partial\bar{\psi}_1}{\partial y_{\rho_1}}\dfrac{\partial a_{\rho_1}}{\partial x} \end{pmatrix}\dfrac{\partial f}{\partial u} \end{bmatrix}.$$

Using (5.16)–(5.19) rewrite (5.11)

$$B^* = \begin{bmatrix} \dfrac{\partial \bar{\tilde{a}}_1}{\partial u} \\ \bar{\alpha}\dfrac{\partial \bar{\tilde{a}}_1}{\partial u} + \alpha_{\rho_1}\dfrac{\partial a_{\rho_1}}{\partial u} \\ \left(\begin{matrix} \dfrac{\partial a_{\rho_1}}{\partial x} + \dfrac{\bar{\alpha}}{\alpha_{\rho_1}}\dfrac{\partial \bar{\tilde{a}}_1}{\partial x} - \dfrac{1}{\alpha_{\rho_1}}\dfrac{\partial a_{\rho_1+1}}{\partial x} \\ \dfrac{\partial \hat{\bar{a}}_1}{\partial x} - \left(\dfrac{\partial \bar{\psi}_1}{\partial \bar{\tilde{y}}_1} + \dfrac{\partial \bar{\psi}_1}{\partial y_{\rho_1}}\dfrac{\bar{\alpha}}{\alpha_{\rho_1}}\right)\dfrac{\partial \bar{\tilde{a}}_1}{\partial x} - \dfrac{\partial \bar{\psi}_1}{\partial y_{\rho_1}}\dfrac{1}{\alpha_{\rho_1}}\dfrac{\partial a_{\rho_1+1}}{\partial x} \end{matrix} \right) \dfrac{\partial f}{\partial u} \end{bmatrix}.$$

Now it is easy to see that

$$\operatorname{rank} B^* = \operatorname{rank} B$$

as

$$B^* = \begin{bmatrix} \begin{matrix} I_{\rho_1-1} & 0 \\ \bar{\alpha} & \alpha_{\rho_1} \end{matrix} & 0 & 0 \\ 0 & -\alpha_{\rho_1} & 0 \\ 0 & -\dfrac{1}{\alpha_{\rho_1}}\dfrac{\partial \bar{\psi}_1}{\partial y_{\rho_1}} & I \end{bmatrix} B\,.$$

This proves the result. ∎

Thus, using the inversion algorithm around a strongly regular equilibrium point, by Lemma 5.2 we obtain a uniquely defined sequence of integers

$$0 \le \rho_1 \le \ldots \le \rho_k \le \ldots \le \min(p, m)\,.$$

Define the rank ρ^* of the system as $\rho^* = \max\{\rho_k, k \ge 1\}$ and let α be defined as the smallest $k \in N$ such that $\rho_k = \rho^*$. In analogy with continuous-time case we call the ρ_k's invertibility indices of the system (5.1). The list of integers ρ_i, $1 \le i \le n$, is called the (algebraic) structure at infinity. These integers form a generalization of the notion of delay orders. The tracking order β of the systm S is defined to be the least integer k such that $\rho_k = p$ or $\beta = \infty$, if $\rho_k < p$ for all $k > 0$.

5.3 Termination of the Inversion Algorithm

In this section we show that around a regular equilibrium point the inversion algorithm terminates in at most n steps, where $n = \dim x$. To be more precise,

(i) ρ_n is the limiting value of a sequence $\{\rho_k, k \ge 1\}$ in the sense that if one were to continue the inversion algorithm, then $\rho_{n+j} = \rho_n$ for all integers $j \ge 0$;

(ii) the number of dependent output components does not decrease after the nth step, that is $\hat{y}_n = \hat{y}_{n+j}$ for all integers $j \geq 0$.

To prove it, we need the following Lemma.

Lemma 5.3 *If the rank of the matrix $\partial A_\alpha(\cdot)/\partial u$ is equal to p (i.e. $\rho_\alpha = p$) in some neighbourhood $\tilde{O}$ of $(x^0, u^0, y^0, \dots y^0) \in X \times U \times Y^{\alpha-1}$ then for $k = 0, 1, \dots, \alpha - 1$ on $\Pi_k(\tilde{O})$ (where $\Pi_k : X \times U \times Y^{\alpha-1} \to X \times Y^k$ is the projection along $U \times Y^{\alpha-1-k}$ on $X \times Y^k$) the following equalities hold:*

$$\operatorname{rank} \frac{\partial}{\partial x} \begin{bmatrix} \psi_0(x) \\ \psi_1(x, \tilde{y}_1(t+1)) \\ \dots \\ \psi_k(x, \{\tilde{y}_i(t+j),\ 1 \leq i \leq k,\ i \leq j \leq k\}) \end{bmatrix} = \sum_{i=0}^{k} (p - \rho_i) . \tag{5.22}$$

Proof. We shall prove the lemma only for the case $k = \alpha - 1$; the proof for the other cases is analogous. Denote by $\psi_{ik}(\cdot)$ the k-th component of $\psi_i(\cdot)$. Assume that

$$\operatorname{rank} \frac{\partial}{\partial x} \begin{bmatrix} \psi_0(\cdot) \\ \dots \\ \psi_{\alpha-1}(\cdot) \end{bmatrix} < \sum_{i=0}^{\alpha-1} (p - \rho_i) ,$$

and let, without loss of generality, the last row of the matrix $\partial \left(\psi_0^T, \dots, \psi_{\alpha-1}^T\right)^T / \partial x$ be linearly dependent on the other rows. Therefore, on a neighbourhood of $(x^0, y^0, \dots \dots, y^0) \in \Pi_{\alpha-1}(\tilde{O})$ there exist functions $\gamma_{ik}(x, \{\tilde{y}_i(t+j),\ 1 \leq i \leq \alpha - 1,\ i \leq j \leq \leq \alpha - 1\})$ such that

$$\frac{\partial}{\partial x} \psi_{\alpha-1, p-\rho_\alpha - 1}(\cdot) = \sum_{i=0}^{\alpha-2} \sum_{k=1}^{p-\rho_i} \gamma_{ik}(\cdot) \frac{\partial}{\partial x} \psi_{ik}(\cdot) + \sum_{k=1}^{p-\rho_{\alpha-1}-1} \gamma_{\alpha-1,k}(\cdot) \frac{\partial}{\partial x} \psi_{\alpha-1,k}(\cdot) . \tag{5.23}$$

Note that (5.23) holds also around the point $(f(x^0, u^0), y^0, \dots, y^0) = = (x^0, y^0, \dots, y^0)$. Multiplying both sides of (5.23) with $\partial f(x,u)/\partial u$ and taking into account that

$$\frac{\partial}{\partial u} \psi_{ij}(f(x,u), \cdot) = \frac{\partial}{\partial x} \psi_{ij}(x, \cdot)_{|x=f(x,u)} \frac{\partial f}{\partial u}_{|(x,u)}$$

we obtain

$$\frac{\partial}{\partial u} \psi_{\alpha-1, p-\rho_{\alpha-1}}(f(x,u), \cdot) =$$

$$= \sum_{i=0}^{\alpha-2} \sum_{k=1}^{p-\rho_i} \gamma_{ik}(\cdot) \frac{\partial}{\partial u} \psi_{ik}(f(x,u), \cdot) + \sum_{k=1}^{p-\rho_{\alpha-1}-1} \gamma_{\alpha-1,k}(\cdot) \frac{\partial}{\partial u} \psi_{\alpha-1,k}(f(x,u), \cdot) .$$

Using the inversion algorithm the last equality results in

$$\frac{\partial}{\partial u}\tilde{y}_{\alpha,p-\rho_{\alpha-1}}(x, \{\tilde{y}_i(t+j),\ 1 \le i \le \alpha-1,\ i \le j \le \alpha-1\}) =$$
$$= \sum_{i=0}^{\alpha-2}\left(\sum_{k=1}^{\rho_{i+1}-\rho_i} \gamma_{ik}(\cdot)\frac{\partial}{\partial u}\tilde{y}_{i+1,k}(\cdot) + \sum_{k=\rho_{i+1}-\rho_i+1}^{p-\rho_i} \gamma_{ik}(\cdot)\frac{\partial}{\partial u}\psi_{i+1,k}(\cdot)\right) +$$
$$+ \sum_{k=1}^{p-\rho_{\alpha-1}-1} \gamma_{\alpha-1,k}(\cdot)\frac{\partial}{\partial u}\tilde{y}_{\alpha,k}(\cdot)\,. \tag{5.24}$$

The left hand side of (5.24) is actually the last row of the matrix $\partial A_\alpha(\cdot)/\partial u$. The only difference is that instead of $\tilde{y}_i(t+j)$ in this matrix there stay $\tilde{y}_i(t+j-1)$ for $1 \le i \le \alpha$, $i+1 \le j \le \alpha$. This fact, of course, does not restrict the generality because $\tilde{y}_i(t+j)$ can take arbitrary values around y^0. Recall that by the inversion algorithm $\partial\psi_{ik}(\cdot)/\partial u = 0$. Thus the right hand side of (5.24) is a linear combination of the $p-1$ first rows of the matrix $\partial A_\alpha(\cdot)/\partial u$ (where $\tilde{y}_i(t+j) = \tilde{y}_i(t+j-1)$). This contradiction proves the lemma. ∎

Next we prove the main result of this section.

Lemma 5.4 *Around a regular equilibrium point the inversion algorithm terminates in at most n steps, i.e.* $\rho^* = \rho_n$.

Proof. Assume at first that $\rho^* = p$. Then by Lemma 5.3 we obtain

$$\operatorname{rank}\frac{\partial}{\partial x}\begin{bmatrix}\psi_0(:)\\ \dots\\ \psi_{\alpha-1}(\cdot)\end{bmatrix} = \sum_{i=0}^{\alpha-1}(p-\rho_i) > \sum_{i=0}^{\alpha-1} 1 = \alpha\,.$$

On the other hand, as $x \in X \subset R^n$, for every x we have

$$\operatorname{rank}\frac{\partial}{\partial x}\begin{bmatrix}\psi_0(\cdot)\\ \dots\\ \psi_{\alpha-1}(\cdot)\end{bmatrix} \le n\,.$$

In the case $\alpha > n$ this will give us a contradiction. Therefore, if $\rho^* = p$, then $\alpha \le n$.

In the case when $\rho^* < p$, we can extract a subsystem from the system (5.1) that has ρ^* outputs, and result still holds. ∎

5.4 Necessary and Sufficient Conditions for Forward Time-shift Right Invertibility

The inversion algorithm incorporates a relatively simple criterion for determining, under the regularity assumption of an equilibrium point, if the system is right invertible.

Theorem 5.5 *Consider the system (5.1) around a regular equilibrium point* (x^0, u^0) *with respect to the inversion algorithm: The system (5.1) is locally around* (x^0, u^0) *forward time-shift right invertible if and only if the rank* ρ^* *of the system is equal to the number of the outputs, that is iff* $\rho^* = p$.

Proof. Sufficiency. If $\rho_\alpha = p$ then at the last α-th step of the inversion algorithm we obtain

$$\begin{aligned} \tilde{y}_1(t+1) &= \tilde{a}_1(x(t), u(t)) \\ \tilde{y}_2(t+2) &= \tilde{a}_2(x(t), u(t), \tilde{y}_1(t+2)) \\ \tilde{y}_3(t+3) &= \tilde{a}_3(x(t), u(t), \tilde{y}_1(t+2), \tilde{y}_1(t+3), \tilde{y}_2(t+3)) \\ &\dots \\ \tilde{y}_\alpha(t+\alpha) &= \tilde{a}_\alpha(x(t), u(t), \{\tilde{y}_i(t+j), i+1 \le j \le \alpha, 1 \le i \le \alpha - 1\}), \end{aligned} \tag{5.25}$$

where at the point (x^0, u^0, y^0) the value of the vector function $A_\alpha(\cdot) = [\tilde{a}_1^T(\cdot), \dots$ $\dots, \tilde{a}_\alpha^T(\cdot)]^T$ is equal to y^0 and moreover, in a neighbourhood of the point (x^0, u^0, y^0)

$$\operatorname{rank} \frac{\partial}{\partial u} A_\alpha(\cdot) = p .$$

Consequently, according to the Implicit Function Theorem the system of equations (5.25) is solvable for the control $u(t)$ around the point (x^0, u^0, y^0), i.e. there exists the analytic function $\varphi(\cdot)$,

$$u(t) = \varphi(x(t), \{\tilde{y}_i(t+j),\ 1 \le i \le \alpha,\ i \le j \le \alpha\}), \tag{5.26}$$

such that $u^0 = \varphi(x^0, y^0, \dots, y^0)$ and around the point (x^0, u^0, y^0) the following holds

$$\begin{aligned} &\left[\tilde{y}_1^T(t+1), \dots, \tilde{y}_\alpha^T(t+\alpha)\right]^T \equiv \\ &\equiv A_\alpha(x(t), \varphi(\cdot), \{\tilde{y}_i(t+j),\ 1 \le i \le \alpha - 1,\ i+1 \le j \le \alpha\}) . \end{aligned} \tag{5.27}$$

Therefore, at $\tilde{y}_i(t)$, $i = 1, \dots, \alpha$, an arbitrary sequence from $\mathcal{Y}_i^0$ is reproducible for $t \ge i$.

Similarly to the case of $(d_1, \dots, d_p)$-forward time-shift right invertibility, the function φ is analytic in some (possible small) neighbourhoods V_3 and V_2 of $(x^0, y^0, \dots, y^0)$ in $X^0 \times (Y^0)^{\alpha-1}$ and u^0 in U^0. It means that for the control (5.24) the equality (5.25) holds as long as $(x(t), \{\tilde{y}_i(t+j),\ 1 \le i \le \alpha,\ i \le j \le \alpha\}) \in V_3$ and $u(t) \in V_2$. The identity (5.25) is lost if we leave the neighbourhoods V_3 or V_2. The sufficiency part has been proved.

Necessity. Suppose that the system (5.1) is locally forward time-shift right invertible around a regular equilibrium point (x^0, u^0). In particular, it means that at $t = i$ as the subvector $\tilde{y}_i(t)$ we can reproduce by suitable choice of $u(0) = u^*$ arbitrary $\tilde{y}_i^*$ sufficiently close to $\tilde{y}_i^0$, i.e. the following holds

$$\begin{aligned} \tilde{y}_1^* &= \tilde{a}_1(x(0), u^*) \\ \tilde{y}_2^* &= \tilde{a}_2(x(0), u^*, \tilde{y}_1(t+2)) \\ &\dots \\ \tilde{y}_\alpha^* &= \tilde{a}_\alpha(x(0), u^*, \{\tilde{y}_i(t+j),\ 1 \le i \le \alpha - 1;\ i+1 \le j \le \alpha\}) . \end{aligned}$$

Assume that the rank of the matrix $\partial A_\alpha(x^0,u^0,y^0,\ldots,y^0)/\partial u = r < p$. As the rank is constant in some small neighbourhood O of the point (x^0,u^0,y^0) by the regularity of the equilibrium point, the rank $\frac{\partial}{\partial u}A_\alpha(\cdot)$ on O is less than p. Consequently, the components of the functions $\tilde{a}_1,\tilde{a}_2,\ldots,\tilde{a}_\alpha$ are functionally dependent in O, i.e. there exists a map R such that

$$R(\tilde{a}_1,\ldots,\tilde{a}_\alpha,x(0)) = R(\tilde{y}_1^*,\ldots,\tilde{y}_\alpha^*,x(0)) = 0$$

which implies that $\tilde{y}_i^*$, $i=1,\ldots,\alpha$ are not arbitrary and gives us a contradiction. The theorem has been proved. ∎

Remark. Clearly $\rho^* = p$ requires $m \geq p$.

Remark. Note that if the system is not invertible, the inversion algorithm provides the maximal number of output components that can be assigned arbitrarily.

5.5 The Construction of Forward Time-shift Right Inverse System

If the inversion algorithm applied to the system S terminates while $\rho^* = p$, that is if the system is right invertible, then the inversion algorithm provides a systematic procedure for constructing a right inverse system.

We are now going to derive the equations of the right inverse system S_R^{-1} for system (5.1), that is the equations of the system, whose I/O map satisfies the equation

$$\begin{aligned}&\text{block diag}\{\delta I_{\rho_1},\delta^2 I_{\rho_2-\rho_1},\ldots,\delta^\alpha I_{\rho_\alpha-\rho_{\alpha-1}}\}\circ\Sigma_S\circ\Sigma_{S_R^{-1}} =\\ &= \text{block diag}\{\delta I_{\rho_1},\delta^2 I_{\rho_2-\rho_1},\ldots,\delta^\alpha I_{\rho_\alpha-\rho_{\alpha-1}}\}.\end{aligned}$$

Note that the derivation is analogous to the case of $(d_1,\ldots,d_p)$-forward time-shift right invertible systems (see Chapter 2). Modify the output equation of the system (5.1) according to the inversion algorithm until we can solve the modified equation for $u(t)$. At the α-th step of the inversion algorithm we reach such a system of equations (5.25). To get the equations for the right inverse, we must solve equations (5.1), (5.25) with respect to $u(t)$ and $x(t+1)$ in terms of $x(t),\tilde{y}_i(t+j)$, $1\leq i\leq\alpha$, $i\leq j\leq\alpha$. The solution of (5.25) is given by (5.26). Equation (5.26) defines the required control of the given system (which yields the reference output) in terms of state, and the future values of the reference output. We can take this equation as the output equation of the right inverse system. To obtain the dynamic part of the right inverse, we must substitute (5.26) into (5.1)

$$x(t+1) = f(x(t),\varphi(x(t),\{\tilde{y}_i(t+j),\ 1\leq i\leq\alpha,\ i\leq j\leq\alpha\}).$$

The resulting inverse

$$\begin{aligned}x^R(t+1) &= f(x^R(t),\varphi(x^R(t),\{\tilde{y}_i(t+j),\ 1\leq i\leq\alpha,\ i\leq j\leq\alpha\}))\\ u(t) &= \varphi(x^R(t),\{\tilde{y}_i(t+j),\ 1\leq i\leq\alpha,\ i\leq j\leq\alpha\}).\end{aligned}\tag{5.28}$$

is not causal.

As in the Chapter 2 we can easily obtain from equations (5.28) the forward time-shift right inverse system in state space form

$$\begin{aligned} \bar{x}^R(t+1) &= f(\bar{x}^R(t), \varphi(\bar{x}^R(t), u^R(t))) \\ y^R(t) &= \varphi(\bar{x}^R(t), u^R(t)), \end{aligned} \tag{5.29}$$

i.e. the system $\bar{S}_R^{-1}$ whose I/O map satisfies the equation

$$\text{block diag}\{\delta I_{\rho_1}, \delta^2 I_{\rho_2-\rho_1}, \ldots, \delta^\alpha I_{\rho_\alpha-\rho_{\alpha-1}}\} \circ \Sigma^S_{x_0} \circ \Sigma^{\bar{S}_R^{-1}}_{\bar{x}_0^R} \circ \begin{bmatrix} \delta[I_{\rho_1} 0] \\ \delta^2[I_{\rho_2} 0] \\ \ldots \\ \delta^\alpha[I_{\rho_\alpha} 0] \end{bmatrix} =$$

$$= \text{block diag}\{\delta I_{\rho_1}, \delta^2 I_{\rho_2-\rho_1}, \ldots, \delta^\alpha I_{\rho_\alpha-\rho_{\alpha-1}}\}.$$

The resulting right inverse system has more inputs than there are outputs in the original system.

Remark. The states of the system (5.1) under feedback $u(t) = \varphi(x(t), u^R(t))$ and the states of the forward time-shift right inverse system (5.29) under the input $u^R(t)$ coincide, provided the initial states of both systems are equal, $x(0) = \bar{x}^R(0)$.

5.6 Reduced-order Right Inverse System. State-free Right Inverse System

In Section 2 it was shown for $(d_1, \ldots, d_p)$-forward time-shift right inverse system that the order of the dynamic part of the inverse system can be reduced to $n - \mu$ (where μ is the sum of delay orders of the given system) without increasing the maximal number of forward time-shifts in the representation of the inverse system. Lemma 5.3 is utilized now to reduce the order of the dynamic part of (5.28), i.e. the inverse system in the general case. It is shown that $r_\alpha = \sum_{i=0}^{\alpha}(p - \rho_i)$ rather than μ plays the role in the generalization.

Theorem 5.6 *Suppose that a forward time-shift right inverse of (5.1) exists and that α is the smallest $k \in N$ such that $\rho_k = p$. Then there exists an inverse system representation whose dynamic part is of order $n - r_\alpha$.*

Proof. Consider the vector valued function $L(x) : R^n \to R^{r_\alpha}$ defined by

$$L(x) = \begin{bmatrix} \psi_0(x) \\ \psi_1(x, \tilde{y}_1(t+1)) \\ \ldots \\ \psi_{\alpha-1}(x, \tilde{y}_i(t+j),\ 1 \le i \le \alpha-1,\ i \le j \le \alpha-1\}) \end{bmatrix}.$$

By Lemma 5.3, the matrix $\partial L(x)/\partial x$ has full row rank r_a. Hence r_a functions $L(x)$ can be completed by $n - r_a$ functions $M(x)$ such that $(\xi, \eta) = (L(x), M(x))$ defines a diffeomorphism on R^n. In new coordinates (ξ, η)

$$[\hat{y}_0^T(t), \hat{y}_1^T(t+1), \ldots, \hat{y}_{\alpha-1}^T(t+\alpha-1)]^T = [\xi_1(t) \ldots \xi_{r_a}(t)]^T .$$

Writing (5.1) in the new coordinates it follows from (5.29) that

$$\begin{aligned} \eta(t+1) &= f_2(\eta(t), u^{RR}(t)) \\ y^R(t) &= \hat{\varphi}(\eta(t), u^{RR}(t)) \end{aligned}$$

where the input of the reduced order inverse is composed of $y(t+j)$, $0 \leq j \leq \alpha$: $u^{RR}(t) = [y^T(t), y^T(t+1), \ldots, y^T(t+\alpha)]^T$. ∎

5.7 Two Versions of the Inversion Algorithm for Systems with Input Disturbances

In this section we shall consider the systems of the form

$$\begin{aligned} x(t+1) &= f(x(t), u(t), w(t)) \\ y(t) &= h(x(t)) \end{aligned} \tag{5.30}$$

where x, u, w, y, f and h are specified as in Chapter 1. For this type of systems two special versions of the inversion algorithm can be considered. The first version accomplishes inversion with respect to both types of inputs, the controls and the disturbances, whereas the other version considers disturbances as system parameters and accomplishes inversion with respect to the controls only. If we define $\nu = (u, w)$, then applying the first version of the IA to system (5.30) means just applying the algorithm given in Section 5.1 with respect to ν. In this section we present the second version of the inversion algorithm for system (5.30).

Step 1. Compute

$$y(t+1) = h(f(x(t), u(t), w(t))) = a_1(x, u, w)$$

and define

$$\rho_{u,1} = \operatorname{rank} \frac{\partial}{\partial u} h(f(x, u, w))\big|_{x=x^0, u=u^0, w=w^0} .$$

Let us assume that rank $\partial h(f(x, u, w))/\partial u$ const in some neighbourhood O_1 of (x^0, u^0, w^0). Permute, if necessary, the components of the output so that the first $\rho_{u,1}$ rows of the matrix $\partial h(f(x, u, w))/\partial u$ are linearly independent. Decompose $y(t+1)$ and $h(f(x, u, w))$ according to

$$y(t+1) = \begin{bmatrix} \tilde{y}_1(t+1) \\ \hat{y}_1(t+1) \end{bmatrix}, \quad h(f(x, u, w)) = \begin{bmatrix} \tilde{a}_1(x, u, w) \\ \hat{a}_1(x, u, w) \end{bmatrix},$$

where $\tilde{y}_1(t+1)$ and $\tilde{a}_1(x, u, w)$ consist of the first $\rho_{u,1}$ components of $y(t+1)$ and $h(f(x, u, w))$ respectively. Since the last $p - \rho_{u,1}$ rows of the matrix $\partial h(f(x, u, w))/\partial u$ are linearly dependent on the first $\rho_{u,1}$ rows, we can write

$$\tilde{y}_1(t+1) = \tilde{a}_1(x(t), u(t), w(t)),$$
$$\hat{y}_1(t+1) = \psi_1(x(t), w(t), \tilde{y}_1(t+1)).$$

Denote $\tilde{a}_1(\cdot)$ by $A_1(\cdot)$.

Step $k+1 (k \geq 1)$. Suppose that in Steps 1 through $k, \tilde{y}_1(t+1), \tilde{y}_2(t+2), \ldots, \tilde{y}_k(t+k), \hat{y}_k(t+k)$ have been defined so that

$$\begin{aligned}
&\tilde{y}_1(t+1) = \tilde{a}_1(x(t), u(t), w(t))\\
&\tilde{y}_2(t+2) = \tilde{a}_2(x(t), u(t), w(t), w(t+1), \tilde{y}_1(t+2))\\
&\quad \ldots\\
&\tilde{y}_k(t+k) = \tilde{a}_k(x(t), u(t), w(t), \ldots, w(t+k-1), \{\tilde{y}_i(t+j), 1 \leq i \leq k-1,\\
&\qquad\qquad i+1 \leq j \leq k\})\\
&\hat{y}_k(t+k) = \psi_k(x(t), w(t), \ldots, w(t+k-1), \{\tilde{y}_i(t+j), 1 \leq i \leq k,\ i \leq j \leq k\}).
\end{aligned}$$

Suppose also that the matrix $\partial A_k(\cdot)/\partial u$, where $A_k = [\tilde{a}_1^T \ldots \tilde{a}_k^T]^T$, has full rank equal to $\rho_{u,k}$ in some neighbourhood O_k of (x^0, u^0, w^0). Compute

$$\begin{aligned}
\hat{y}_k(t+k+1) &= \psi_k(f(x(t), u(t), w(t)), w(t+1), \ldots\\
&\qquad \ldots, w(t+k), \{\tilde{y}_i(t+j+1),\ 1 \leq i \leq k,\ i \leq j \leq k\}) =\\
&= a_{k+1}(x(t), u(t), w(t), \ldots\\
&\qquad \ldots, w(t+k), \{\tilde{y}_i(t+j),\ 1 \leq i \leq k,\ i+1 \leq j \leq k+1\})
\end{aligned}$$

and define

$$\rho_{u,k+1} = \operatorname{rank} \frac{\partial}{\partial u} \begin{bmatrix} A_k(\cdot) \\ a_{k+1}(\cdot) \end{bmatrix}_{x=x^0, u=u^0, w=w^0}.$$

Let us assume that rank $\partial[A_k^T, a_{k+1}^T]^T/\partial u$ const in some neighbourhood O_{k+1} of (x^0, u^0, w^0). Permute, if necessary, the components of $\hat{y}_k(t+k+1)$ so that the first $\rho_{u,k+1}$ rows of the matrix $\partial[A_k^T, a_{k+1}^T]^T/\partial u$ are linearly independent. Decompose $\hat{y}_k(t+k+1)$ and a_{k+1} according to

$$\hat{y}_k(t+k+1) = \begin{bmatrix} \tilde{y}_{k+1}(t+k+1) \\ \hat{y}_{k+1}(t+k+1) \end{bmatrix}, \quad a_{k+1} = \begin{bmatrix} \tilde{a}_{k+1} \\ \hat{a}_{k+1} \end{bmatrix},$$

where $\tilde{y}_{k+1}(t+k+1)$ and $\tilde{a}_{k+1}$ consist of the first $\rho_{u,k+1} - \rho_{u,k}$ components of $\hat{y}_k(t+k+1)$ and a_{k+1} respectively. Since the last $p - \rho_{u,k+1}$ rows of the matrix $\partial[A_k^T, a_{k+1}^T]^T/\partial u$ are linearly dependent on the first $\rho_{u,k+1}$ rows, we can write

$$\begin{aligned}
\tilde{y}_1(t+1) &= \tilde{a}_1(x(t), u(t), w(t))\\
&\ldots\\
\tilde{y}_{k+1}(t+k+1) &= \tilde{a}_{k+1}(x(t), u(t), w(t), \ldots, w(t+k),\\
&\qquad \{\tilde{y}_i(t+j), 1 \leq i \leq k,\ i+1 \leq j \leq k+1\})\\
\hat{y}_{k+1}(t+k+1) &= \psi_{k+1}(x(t), w(t), \ldots, w(t+k),\\
&\qquad \{\tilde{y}_i(t+j), 1 \leq i \leq k+1,\ i \leq j \leq k+1\})
\end{aligned}$$

Denote $A_{k+1} = [A_k^T, \tilde{a}_{k+1}^T]^T$. End of the step $k+1$.

In the sequel a notion of regularity, associated with the second version of the inversion algorithm, of an equilibrium point will be defined.

Definition 5.7 *We call the equilibrium point (x^0, u^0, w^0) of the system (5.30) regular with respect to the second version of the IA if for some specific application i of the inversion algorithm the constant rank assumptions of the algorithm are satisfied. We call (x^0, u^0, w^0) strongly regular if the constant rank assumptions of the algorithm hold for each application of the algorithm.*

Using the second version of the inversion algorithm around a regular equilibrium point (x^0, u^0, w^0) of the system (5.30), we obtain a sequence of integers

$$0 \leq \rho_{u,1} \leq \ldots \leq \rho_{u,k} \leq \ldots \leq p .$$

Let $\rho_u^* = \max\{\rho_{u,k}, k \geq 1\}$ and let α be defined as the smallest $k \in N$ such that $\rho_{u,k} = \rho_u^*$. The result of the inversion algorithm, that is the functions $\tilde{a}_i(\cdot), i \geq 1$ and $\psi_i(\cdot), i \geq 1$, is not unique: it depends on the choice of admissible permutation of output components at every step of the algorithm. In analogy with the case without input disturbances it can be proved that the integers $\rho_{u,1}, \ldots, \rho_{u,k}, \ldots$ do not depend on the particular permutation of the components of $\hat{y}_k(t+k+1)$. Thus around a strongly regular equilibrium point of the system, these integers define some structural properties of the system. We call the $\rho_{u,k}, k \geq 1$ the invertibility indices of the system (5.30) with respect to control.

The sequence of integers $0 \leq \rho_{uw,1} \leq \ldots \leq \rho_{uw,k} \leq \ldots \leq p$ we obtain when using the first version of the IA, will be called the invertibility indices with respect to both inputs, the controls and the disturbances.

Moreover, in analogy with the case without disturbances it can be proved that around a regular equilibrium point the inversion algorithm terminates in at most n steps, that is

$$\rho_u^* = \rho_{u,n} .$$

Notes and References

The inversion algorithm [Sil69], [Sin81], [LF87], [LFG93] has been generalized for discrete-time nonlinear systems in [Kot90]. The more efficient form which corresponds to the one presented in Section 5.1, has been given in [KN91] and is actually a discrete-time analogue of [BGM89]. A restricted version of the inversion algorithm has been introduced by Lee and Marcus in [LM87]. In order to apply the resticted version of the algorithm, the system must satisfy certain additional assumptions. Namely, the function $\psi_k(\cdot)$, $k \geq 1$, obtained at the kth step of the algorithm, must have the following form

$$\psi_k(x(t), \{\tilde{y}_i(t+j), 1 \leq i \leq k, i \leq j \leq k\}) = \sum_{i=1}^{k} \sum_{j=i}^{k} K_{ij}^k \tilde{y}_i(t+j) + b_k(x(t)) . \quad (5.31)$$

The inversion algorithm, presented in Section 5.1, involves the solution of system of nonlinear equations and therefore, in general, asks for the use of the Implicit Function Theorem. Recently, Grizzle [Gri93] introduced another form of the inversion algorithm which, instead of working with the outputs $y(t)$, works with differentials $dy(t)$ of the output, thus linearizing the computations. A new form of the algorithm is certainly more efficient for computing the invertibility indices ρ_k and the rank ρ^* of the system. Of course, this form of the algorithm cannot be used to find the equations of the right inverse system. For this purpose, we still need the inversion algorithm, presented in Section 5.1.

Our approach to the problem of right invertibility is based on the inversion algorithm which provides both the criterion for determining if the system is right invertible and a systematic procedure for constructing a right inverse system if the system is right invertible. Another way of achieving a constructive procedure for the local right invertibility is given in [Nij90]. The difference is that, in the inversion algorithms [LM87], [Kot90], [KN91], [Gri93], state dependent transformations of the outputs are used, whereas in the so-called extension algorithm [Nij87], state dependent transformations of the inputs (the static-state feedback) are used. The latter enables to make a direct connection with the input-output decoupling problem.

If it is possible to carry out the inversion algorithm or its dual variant, the extension algorithm, then the system is right invertible. The important point to emphasize is that in order to apply the above algorithms, certain constant rank conditions at each successive step of the algorithm have to hold. These conditions were formalized in the notion of regularity of the equilibrium point, associated with the inversion algorithm (see Sec. 5.1). Our definition of regularity of the equilibrium point reminds the one of [Hui91] in continuous-time. In case the equilibrium point is not regular, the inversion method in the present form breaks down. However, the system can be invertible without being able to carry out the inversion algorithm. It was recently showed (in case of continuous-time nonlinear systems) by Respondek [Res93] that the inversion algorithm-based regularity conditions can be softened. He introduced a new regularity notion consisting also in a constant rank requirement but only after a certain order of application of shift operator δ on the outputs, thus allowing some rank degeneracies before that order. The idea of softening the inversion algorithm-based regularity conditions was first proposed for linear time-varying systems by Grasse [Gra88]; see also [Cam89]. Grasse conjectured the necessity of certain rank conditions and showed that the system may be right invertible even if the inversion algorithm cannot be carried out because of time-varying ranks at intermediate steps.

A different approach for studying invertibility of discrete-time nonlinear systems is based upon difference algebra [EAF92] where the invertibility is defined in terms of transformal transcendence degree of the system. The latter gives the maximal number of the output components which are algebraically independent, i.e. can not be related by any algebraic difference equation.

Linear algebraic framework was extended to the class of discrete-time nonlinear systems by Grizzle [Gri93]. Through the introduction of a chain of subspaces naturally associated with the output of a system, it provides a system-intrinsic interpretation of the inversion algorithm, showing that the invertibility indices defined by the inversion algorihm are actually tied with dimensions of certain subspaces, providing so algorithm-independent structural parameters of the system. In addition, via the linear algebraic approach the relationship between the inversion algorithm and the difference algebraic approach, is established. In [Gri93] it has been proved that the rank ρ^* defined by the inversion algorithm, when specialized to systems whose right hand side depends polynomially on components of x and u, corresponds to the transformal transcendence degree.

It has been proved by Kotta and Nijmeijer [KN91] that the integers ρ_k, $k \geq 0$, do not depend on the particular application of the inversion algorithm. The above integers are called the invertibility integers in analogy with continuous-time case [Moo88]. In [GN86] another set of integer invariants for a discrete-time nonlinear system has been introduced, the so-called (geometric) stucture at infinity. For linear systems the both sets coincide, this is not the case for nonlinear systems. The geometric structure at infinity has some rather surprising anomalies [GN86] if compared to the linear case. For instance, the number of zeros at infinity can exceed the number of outputs, which is of course impossible for a linear system. Next, in the case of linear systems, zeros at infinity can be used to characterize when the system is right invertible. Indeed, a linear system is right invertible if and only if the number of zeros at infinity equals the number of outputs. Unfortunately, this does not hold for general nonlinear discrete-time systems. The algebraic structure at infinity does not have the above anomalies and can be considered the more appropriate generalization of the structure at infinity to the nonlinear case. However, for the subclass of systems which satisfy (5.31) (i.e. for the systems to which the special case of inversion algorithm by Lee and Marcus [LM87] is applicable) the geometric and algebraic structures at infinity coincide.

The output zeroing problem and the zero dynamics algorithm are closely related to the problem of right invertibility and the inversion algorithm, respectively. The zero dynamics algorithm has been generalized in [MNC87] for discrete-time nonlinear systems; see also [Kot93] for systems with measurable input disturbances.

The results described in Sections 5.4 and 5.5 may be found in [Kot90]. The Lemma 5.3 and 5.4 have been proved in [Kot92]. The reduced-order and state-free (left) inverse systems have been constructed in [MNC87].

The idea of using two versions of inversion algorithm [MPC91] was extended to discrete-time nonlinear systems in [Kot92a].

[BGM89] Di Benetto M.D., J.W.Grizzle and C.H.Moog. Rank invariants of nonlinear systems. *SIAM J. Control and Optimiz.*, 1989, v. 27, 658–672.

[Cam89] Campbell S.L. Comments on functional reproducibility of time-varying input-output systems. *IEEE Trans. on Automatic Control*, 1989, v. 34, 922–924.

[EAF92] El Asmi S. and M.Fliess. Invertibility of discrete-time systems. *Proc. 2nd IFAC Symp. on Nonlinear Control Systems Design*, Bordeaux, 1992, 192–196.

[Gra88] Grasse K.A. Sufficient conditions for the functional reproducibility of time-varying, input-output systems. *SIAM J. Control and Optimization*, 1988, v. 26, 230–249.

[Gri93] Grizzle J.W. A linear algebraic framework for the analysis of discrete-time nonlinear systems. *SIAM J. Control Optimiz.*, 1993, v. 31, 1026–1044.

[GN86] Grizzle J.W. and H.Nijmeijer. Zeros at infinity for nonlinear discrete time systems. *Math. Systems Theory*, 1986, v. 19, 79–93.

[Hui91] Huijberts H.J.C. Dynamic Feedback in Nonlinear Synthesis Problems. Ph.D. Thesis. University of Twente, Enschede, 1991.

[Kot90] Kotta Ü. Right inverse of a discrete-time non-linear system. *Int. J. Control*, 1990, v. 51, 1–9.

[Kot92a] Kotta Ü. Model matching of nonlinear discrete-time systems in the presence of unmeasurable disturbance. *Prepr. IFAC Symp. on Nonlinear Control Systems Design*, Bordeaux, 1992, 563–568.

[Kot92b] Kotta Ü. Dynamic disturbance decoupling for discrete-time nonlinear systems: the nonsquare and noninvertible case. *Proc. Estonian Acad. Sci. Phys. Math.*, 1992, v. 41, 14–22.

[Kot93] Kotta Ü. On zero dynamics of discrete-time nonlinear system in the presence of disturbances. *Prepr. of European Control Conference*, Groningen, 1993, v. 2, 767–769.

[KN91] Kotta Ü. and H.Nijmeijer. Dynamic disturbance decoupling for nonlinear discrete-time systems (In Russian). *Proc. of the Academy of Sciences of USSR. Technical Cybernetics*, 1991, 52–59.

[LF87] Li S.-W. and Y.-K.Feng. Functional reproducibility of general multivariable analytic nonlinear systems. *Int. J. Control*, 1987, v. 45, 255–268.

[LFG93] Li S.-W., Y.-K.Feng and Y.Gong. Comment on 'Functional reproducibility of general multivariable analytic nonlinear systems'. *Int. J. Contr.*, 1993, v. 58, 733–734.

[LM87] Lee H.G. and S.I.Marcus. On input-output linearization of discrete-time nonlinear systems. *Systems and Control Letters*, 1987, v. 8, 249–259.

[MNC87] Monaco S. and D.Normand-Cyrot. Minimum-phase nonlinear discrete-time systems and feedback stabilization. *Proc. 26th IEEE Conf. on Decision and Control*, Los Angeles, CA, 1987, 979–986.

[Moo88] Moog C.H. Nonlinear decoupling and structure at infinity. *Math. of Contr., Signals and Syst.*, 1988, v. 1, 257–268.

[MPC91] Moog C.H., A.M.Perdon and G.Conte. Model matching and factorization for nonlinear systems: a structural approach. *SIAM J. Control and Optimization*, 1991, v. 29, 769–785.

[Nij87] Nijmeijer H. Local (dynamic) input-output decoupling of discrete-time nonlinear systems. *IMA J. Math. Contr., Inform.*, 1987, v. 4, 237–250.

[Nij90] Nijmeijer H. Remarks on the control of discrete-time nonlinear systems. In: *Perspectives in Control Theory.* (Eds. B.Jakubczyk, K.Malanowski, W.Respondek). Birkhäuser, Boston Inc., 1990, 261–276.

[Res93] Respondek W. Dynamic input-output linearization and decoupling of nonlinear systems. *Proc. European Control Conference*, Groningen, 1993.

[Sil69] Silverman L.M. Inversion of multivariable linear systems. *IEEE Trans. Autom. Contr.*, 1969, v. 14, 270–276.

[Sin81] Singh S.N. A modified algorithm for invertibility in nonlinear systems. *IEEE Trans. Autom. Contr.*, 1981, v. 26, 595–598.

6. Applications of the Inversion Method

6.1 The Solution of the Model Matching Problem

This section considers the problem of designing a compensator for a nonlinear discrete-time system S under which input-output behavior of the compensated system becomes the same as the one of the prespecified nonlinear model M. In Chapter 3 we posed and solved the problem under the additional assumption that the system S is $(d_1, \ldots, d_p)$-forward time-shift right invertible. In this Chapter we study the general case.

Definition 6.1 Local model matching problem. *Given the system S defined by equations (5.1) around an equilibrium point (x^0, u^0), the model M defined by equations (3.2) around an equilibrium point (x^{M0}, u^{M0}) corresponding to (x^0, u^0) and a point $(x(0), x^M(0))$, find, if possible, a compensator C defined by equations of the form (3.3) together with an initial state $x^C(0)$, an equilibrium point x^{C0}, neighbourhoods $V_1 = X^0 \times X^{C0} \times U^{M0}$ of (x^0, x^{C0}, u^{M0}) in $X \times X^C \times U^M$ and V_2 of u^0 in U, being the domain and the range of C respectively, as well as neighbourhood X^{M0} of x^{M0} and a map $\xi : X^{M0} \to X^{C0}$ with the property that*

$$\begin{aligned} y^{S\circ C}(t, x(0), \xi(x^M(0)), u^M(0), \ldots, u^M(t-1)) - \\ - y^M(t, x^M(0), u^M(0), \ldots, u^M(t-1)),\ 0 \le t \le t_F \end{aligned}$$

does not depend on u^M for all $(x(0), x^M(0)) \in X^0 \times X^{M0}$ and for all u^M in the domain of C.

The solution is obtained via the inversion (structure) algorithm and is formulated in terms of the invertibility indices of the so-called extended system SM, formed by the original system S and the model M.

An extended system SM of the form (1.1) can be associated with the original system S and the model M as follows:

$$\begin{aligned} x(t+1) &= f(x(t), u(t)) \\ x^M(t+1) &= f^M(x^M(t), u^M(t)) \\ y^{SM}(t) &= h^{SM}(x(t), x^M(t)) = h(x(t)) - h^M(x^M(t)) \end{aligned} \tag{6.1}$$

with the state $x^{SM} = (x^T, x^{M,T})^T$, the control $u(t)$, the measurable disturbances $u^M(t)$, the output $y^{SM}(t)$ and with the equilibrium point $(x^0, x^{M0}, u^0, u^{M0})$.

In the proof of our main result we need the following Lemma.

Lemma 6.2 *Around a strongly regular equilibrium point $(x^0, x^{M0}, u^0, u^{M0})$ of SM the invertibility indices with respect to the control of the extended system SM and the original system S are equal, i.e.*

$$\rho_{u,i}(SM) = \rho_{u,i}(S)\,,\ 1 \geq 1\,.$$

In particular, the ranks with respect to the control of both systems are equal, i.e. $\rho_u^*(SM) = \rho_u^*(S)$.

Proof. The proof is straightforward. Applying the first step of the inversion algorithm (with respect to u) for SM, we obtain

$$\begin{aligned}\rho_{u,1}(SM) &= \text{rank}\frac{\partial}{\partial u}[h(f(x(t),u(t))) - h^M(f^M(x^M(t),u^M(t)))] = \\ &= \text{rank}\frac{\partial}{\partial u}h(f(x(t),u(t))) = \rho_{u,1}(S)\,.\end{aligned}$$

From the special form of the output function h^{SM} we have

$$\begin{aligned}\tilde{y}_1^{SM}(t+1) &= \tilde{a}_1^S(x(t),u(t)) - \tilde{a}_1^*(x^M(t),u^M(t)) = \\ &= \tilde{y}_1^S(t+1) - \tilde{a}_1^*(x^M(t),u^M(t)) \\ \hat{y}_1^{SM}(t+1) &= \hat{a}_1^S(x(t),u(t)) - \hat{a}_1^*(x^M(t),u^M(t)) = \\ &= \psi_1^S(x(t),\tilde{y}_1^S(t+1)) - \hat{a}_1^*(x^M(t),u^M(t))\,.\end{aligned}$$

At the second step of the inversion algorithm we compute

$$\begin{aligned}\hat{y}_1^{SM}(t+2) &= \psi_1^S(x(t+1),\tilde{y}_1^S(t+2)) - \hat{a}_1^*(x^M(t+1),u^M(t+1)) = \\ &= a_2^S(x(t),u(t),\tilde{y}_1^S(t+2)) - a_2^*(x^M(t),u^M(t),u^M(t+1))\end{aligned}$$

and define

$$\rho_{u,2}(SM) = \text{rank}\,\frac{\partial}{\partial u}\begin{bmatrix}\tilde{a}_1^S(\cdot) - \tilde{a}_1^*(\cdot)\\ a_2^S(\cdot) - a_2^*(\cdot)\end{bmatrix} = \text{rank}\,\frac{\partial}{\partial u}\begin{bmatrix}\tilde{a}_1^S(\cdot)\\ a_2^S(\cdot)\end{bmatrix} = \rho_{u,2}(S)\,.$$

Decomposing $\hat{y}_1^{SM}(t+2)$ and $a_2^S(\cdot) - a_2^*(\cdot)$, we have:

$$\begin{aligned}\tilde{y}_2^{SM}(t+2) &= \tilde{a}_2^S(x(t),u(t),\tilde{y}_1^S(t+2)) - \tilde{a}_2^*(x^M(t),u^M(t),u^M(t+1)) = \\ &= \tilde{y}_2^S(t+2) - \tilde{a}_2^*(x^M(t),u^M(t),u^M(t+1)) \\ \hat{y}_2^{SM}(t+2) &= \psi_2^S(x(t),\tilde{y}_1^S(t+1),\tilde{y}_1^S(t+2),\tilde{y}_2^S(t+2)) - \\ &\quad -\tilde{a}_2^*(x^M(t),u^M(t),u^M(t+1))\,.\end{aligned}$$

The rest of the proof runs as before. ∎

Theorem 6.3 *Consider the system S described by equations (3.1) around an equilibrium point (x^0, u^0) and the model M described by equations (3.2) around an equilibrium point (x^{M0}, u^{M0}), corresponding to (x^0, u^0). Assume that the equilibrium point*

$(x^0, x^{M0}, u^0, u^{M0})$ of the extended system SM is strongly regular with respect to both versions of the inversion algorithm. The local model matching problem for S and M is solvable if and only if the invertibility indices of the extended system SM and the original system S are equal, i.e. if and only if

$$\rho_{uu^M,i}(SM) = \rho_{u,i}(S)\,,\ i \geq 1\,. \tag{6.2}$$

Proof. Sufficiency. Assume that (6.2) holds. Then by Lemma 6.2 we have

$$\rho_{u,i}(SM) = \rho_{uu^M,i}(SM)\,. \tag{6.3}$$

which implies that both versions of the inversion algorithm applied to system SM coincide and consequently, at the αth step of the algorithm we obtain

$$\begin{aligned}&[\tilde{y}_1^{SM,T}(t+1),\ldots,\tilde{y}_\alpha^{SM,T}(t+\alpha)]^T = \\ &\quad = A_\alpha(x(t), x^M(t), u(t), u^M(t), \{\tilde{y}_i^{SM}(t+j)\,,\ 1 \leq i \leq \alpha-1\,,\ i+1 \leq j \leq \alpha\})\end{aligned} \tag{6.4}$$

and

$$\hat{y}_\alpha^{SM}(t+\alpha) = \psi_\alpha^{SM}(x(t), x^M(t), \{\tilde{y}_i^{SM}(t+j)\,,\ 1 \leq i \leq \alpha\,,\ i \leq j \leq \alpha\})\,. \tag{6.5}$$

According to the inversion algorithm the Jacobian matrix of the right hand side of (6.4) with respect to u has full row rank ρ_u^* around the regular equilibrium point $(x^0, x^{M0}, u^0, u^{M0})$. Moreover, $A_\alpha(x^0, x^{M0}, u^0, u^{M0}, 0, \ldots, 0) = 0$. So we may solve equation (6.4) for $u(t)$ (not necessarily uniquely) around the equilibrium point $(x^0, x^{M0}, u^0, u^{M0})$ by applying the Implicit Function Theorem. Without loss of generality we can choose zero values to $\tilde{y}_i^{SM}(t+j)$ in (6.4). In this way we obtain a solution of (6.4)

$$u(t) = \varphi(x(t), x^M(t), u^M(t)) \tag{6.6}$$

which is such that

$$A_\alpha(x(t), x^M(t), \varphi(x(t), x^M(t), u^M(t)), u^M(t), 0, \ldots, 0) \equiv 0\,. \tag{6.7}$$

Notice that $\varphi : V_1 \to V_2$ is defined for some (possible small) neighbourhoods V_1 and V_2 of (x^0, x^{M0}, u^{M0}) in $X \times X^M \times U^M$ and u^0 in U. The equality (6.7) is lost if we leave either V_1 or V_2.

Construct the compensator C as

$$\begin{aligned}x^C(t+1) &= f^M(x^C(t), u^M(t))\,,\ x^C(0) = x_0^M\,,\\ u(t) &= \varphi(x(t), x^C(t), u^M(t))\,.\end{aligned} \tag{6.8}$$

We claim that (6.8) and $\xi = \mathrm{Id}$ (identity map) serve as a solution of the MMP. In order to show this, let us first remark that by (6.3), $\hat{y}_\alpha^{SM}(t+k)$ in (6.5) remains independent from $u^M(t)$ for all $k > \alpha$, since otherwise $\rho_{uu^M,k}(SM)$ would be strictly greater than $\rho_{u,k}(SM)$. Therefore, $\hat{y}_k^{SoC}(t+k) - \hat{y}_k^M(t+k) = \hat{y}_k^{SM}(t+k)$ is independent from u^M for every $k \geq 1$. From (6.7), it follows that $\tilde{y}_j^{SoC}(t+j) = \tilde{y}_j^M(t+j)$,

$j = 1, \ldots, \alpha$. So, $y^{SoC}(t) - y^M(t)$ is independent from u^M for $0 \le t \le t_F$. This completes the sufficiency part of the proof.

Necessity. Suppose that there exists a compensator C of the form (3.3) for S and M that, around the strongly regular equilibrium point $(x^0, x^{M0}, u^0, u^{M0})$ of SM with respect to both versions of the inversion algorithm, locally solves the MMP. Apply the first step of the inversion algorithm to SM with respect to control u only, considering disturbances u^M as parameters, to obtain

$$\begin{aligned}\tilde{y}_1^{SM}(t+1) &= \tilde{a}_1^{SM}(x(t), x^M(t), u(t), u^M(t)) \\ \hat{y}_1^{SM}(t+1) &= \psi_1^{SM}(x(t), x^M(t), u^M(t), \tilde{y}_1^M(t+1)),\end{aligned} \tag{6.9}$$

where by the inversion algorithm rank $\partial \tilde{a}_1^{SM}(\cdot)/\partial u = \rho_{u,1}(SM)$.

If we plug the output of C in (6.9), the equations no longer depend on u^M, since C solves the MMP for S and M. In particular this means that either

$$\partial \psi_1^{SM}(\cdot)/\partial u^M = \gamma(x, x^M, u^M, \tilde{y}_1^M) \equiv 0 \tag{6.10}$$

everywhere around the point $(x^0, x^{M0}, u^{M0}, y^{SM,0})$, or the compensator C will guarantee the equality (6.10). Note that around the strongly regular equilibrium point $\partial \psi_1^{SM}(\cdot)/\partial u^M$ is everywhere either equal to zero or different from zero since otherwise $\rho_{uu^M,1}(SM)$ would not be constant around the equilibrium point. This means that working around a strongly regular equilibrium point we can never make $\partial \psi_1^{SM}(\cdot)/\partial u^M$ equal to zero by a suitable choice of the compensator since doing so we fall into the nonregular equilibrium point. According to the above remark, we have $\partial \psi_1^{SM}(\cdot)/\partial u^M \equiv 0$ which gives us

$$\rho_{uu^M,1}(SM) = \operatorname{rank}\begin{bmatrix} \partial \tilde{a}_1^{SM}/\partial u & \partial \tilde{a}_1^{SM}/\partial u^M \\ 0 & \partial \psi_1^{SM}/\partial u^M \end{bmatrix} = \operatorname{rank}\frac{\partial}{\partial u}\tilde{a}_1^{SM} = \rho_{u,1}(SM).$$

Applying this argument repeatedly, we finally obtain

$$\rho_{uu^M,1}(SM) = \rho_{u,i}(SM), \; i \ge 1.$$

The conclusion of the necessity part of the theorem follows using Lemma 6.2. ∎

Remark. Of course, working around a nonregular equilibrium point associated with the inversion algorithm with respect to both inputs it is sometimes possible to make $\partial \psi_k^{SM}/\partial u^M$ equal to zero by the proper choice of C. Such a choice will exploit the nonregularity of the equilibrium point around which we work, namely with this choice we fall into a nonregular equilibrium point. Note that $\rho_{uu^M,k}(SM)$, evaluated at this point, is less than optimal, and exactly equal to $\rho_{u,k}(SM)$.

The following example will illustrate the above remark

Example 6.4 Consider the system S, described as

$$\begin{aligned} x_1(t+1) &= u_1(t) \\ x_2(t+1) &= u_2(t) \\ x_3(t+1) &= x_1(t) + x_1(t)u_2(t) \\ y_1(t) &= x_1(t), \; y_2(t) = x_2(t), \; y_3(t) = x_3(t) \end{aligned}$$

and the model M, described by the equations

$$\begin{aligned}
x_1^M(t+1) &= x_4^M(t) + u_1^M(t)\\
x_2^M(t+1) &= x_6^M(t)\\
x_3^M(t+1) &= x_5^M(t)\\
x_4^M(t+1) &= u_2^M(t)\\
x_5^M(t+1) &= x_4^M(t) + u_1^M(t) + x_6^M(t)\\
x_6^M(t+1) &= x_6^M(t)\\
y_1^M(t) = x_1^M(t)\,,\ y_2^M(t) &= x_2^M(t)\,,\ y_3^M(t) = x_3^M(t)\,.
\end{aligned}$$

The compensator, described by

$$\begin{aligned}
x^C(t+1) &= u_2^M(t)\\
u_1(t) &= x^C(t) + u_1^M(t)\\
u_2(t) &= 0
\end{aligned}$$

solves the MMP, although the conditions (6.2) do not hold for $k = 2$. Indeed, $\rho_{u,2}(S) = 2$, but

$$\rho_{uu^M,2}(SM) = \operatorname{rank}\begin{pmatrix} 1 & 0 & -1\\ 0 & 1 & 0\\ 1 + x_2(t+2) & 0 & -1 \end{pmatrix}$$

is not constant around the equilibrium point $(0,0,0,0)$ of the extended system: in case $x_2 = 0$ it is equal to two, but in all other cases it is equal to three.

Around a strongly regular equilibrium point of the extended system SM with respect to both versions of the inversion algorithm, necessary and sufficient conditions (6.2) for solvability of the nonlinear discrete-time MMP are in full accordance with the corresponding conditions for linear systems. Of course, this similarity no longer holds around a nonregular equilibrium point, where the conditions (6.2) are not necessary.

Note that the assumption of regularity of the equilibrium point of the extended system SM with respect to both versions of the inversion algorithm is slightly stronger than the assumption actually needed in the sufficiency part of the proof of Theorem 6.3. The latter requires regularity of the equilibrium point with respect to the first version (inversion with respect to u) of IA only.

Remark. The compensator (6.8) is not the only possible compensator that solves the MMP for S and M. Recall that we obtained (6.8) by solving equation (6.4) under the assumption that $\tilde{y}_i^{SM}(t+j) = 0$. We may generalize the procedure for obtaining the MMP solving compensator in the following way. We may assume (if it is possible in case of fixed S and M) that the vector functions $\tilde{y}_i^{SM}(t+j)$, $1 \le i \le \alpha$, $i \le j \le \alpha$, $0 \le t+j \le t_F$ have a form $\Phi_{ij}(x(t), x^M(t))$, where

1) Φ_{ij} is a function which for all $k \ge 0$ satisfies the following relation for some function Φ_{ij}^k

$$\Phi_{ij}(x(t+k), x^M(t+k)) = \Phi_i^k(x(t), x^M(t)),$$

i.e. that $\Phi_{ij}(x(t+k), x^M(t+k))$ does not depend on control $u(t)$ and $u^M(t)$.

2) $\Phi_{ij}(x^0, x^{M0}) = 0$.

Note however, that if in case of the compensator (6.8) the difference between the outputs of the closed loop system and the model will be equal to zero beginning from some time instant, then in case of the generalized compensator

$$\tilde{y}_i^{SoC}(t+i) - \tilde{y}_i^M(t+i) = \Phi_i^t(x(t), x^M(t)), \; 1 \le i \le \alpha, \; 0 \le t+i \le t_F.$$

Corollary 6.5 *Consider the system S described by equations (3.1) around an equilibrium point (x^0, u^0) and the model M described by equations (3.2) around an equilibrium point (x^{M0}, u^{M0}). Assume that the system is right invertible around (x^0, u^0). Then the local model matching problem for S and M is solvable if the delay orders of the model are greater than the corresponding essential orders of the system, that is*

$$d_i(M) \ge \varepsilon_i(S), \; i = 1, \ldots, p.$$

6.2 The Solution of the Input-Output Linearization Problem via Static State Feedback

In this section, like in Section 3.3, we study the problem of linearizing the input-output map of a nonlinear system via static state feedback locally around a regular equilibrium point. But unlike Section 3.3 where the problem has been solved for $(d_1, \ldots, d_p)$-forward time-shift right invertible systems, we consider now the general case where no additional assumptions have been made on system S. Necessary and sufficient conditions are given for the existence of a regular static state feedback control law under which the input-dependent part of the response of a nonlinear system becomes linear in the input and independent of the initial state.

So, in this Section we are looking for an analytic static state feedback with a new m-dimensional control v, described by equations of the form

$$u(t) = \varphi(x(t), v(t)), \tag{6.11}$$

defined locally around a point (x^0, v^0, u^0) where v^0 may be found from the equality

$$u^0 = \varphi(x^0, v^0).$$

We call the feedback described by equation (6.11) regular if $\partial\varphi(\cdot)/\partial v$ is nonsingular around (x^0, v^0), that is, if the map $\varphi(x, \cdot) : V^0 \to U^0$ is invertible around the point (x^0, v^0, u^0).

The closed-loop system (3.1), (6.11), initialized at x_0, that is the system

$$
\begin{aligned}
x(t+1) &= f(x(t), \varphi(x(t), v(t))), \; x(0) = x_0, \\
y(t) &= h(x(t))
\end{aligned}
\tag{6.12}
$$

is denoted by $S \circ C$.

The closed-loop system (6.12) is said to have a linear input-output map if its Volterra series expansion reduces to one of the form

$$
y(t) = w^0(t, x_0) + \sum_{i=1}^{m} \sum_{\tau=0}^{t-1} w_i^1(t-\tau)[v_i(\tau) - v_i^0], \; 0 \le t \le t_F. \tag{6.13}
$$

Definition 6.6 Local static input-output linearization problem. *Given the system (3.1) around an equilibrium point (x^0, u^0, y^0) find, if possible, a regular static state feedback C defined by equation of the form (6.11), neighbourhoods $\Lambda = X^0 \times V^0$ of (x^0, v^0) and U^0 of u^0, so that the outputs of the closed-loop system*

$$
y^{S \circ C}(t, x_0, v(0), \ldots, v(t-1)) = w^0(t, x_0) + \sum_{i=1}^{m} \sum_{\tau=0}^{t-1} w_i^1(t-\tau)[v_i(\tau) - v_i^0], \; 0 \le t \le t_F,
$$

for every $x_0 \in X^0$, $v(k) \in V^0$, $0 \le k \le t_F - 1$.

The solution of the considered problem is formulated in terms of the inversion algorithm.

Theorem 6.7 *Consider the system S described by equations (3.1) around a regular equilibrium point with respect to the inversion algorithm. The static input-output linearization problem for S is locally solvable around the regular equilibrium point if and only if the functions*

$$
\psi_k(x(t), \{\tilde{y}_i(t+j), \; 1 \le i \le k, \; i \le j \le k\}), \; k \ge 1
$$

in some application of the inversion algorithm (See (5.4)) reduce to the following special form

$$
\psi_k(x(t), \{\tilde{y}_i(t+j), \; 1 \le i \le k, \; i \le j \le k\}) = \sum_{i=1}^{k} \sum_{j=i}^{k} K_{ij}^k \tilde{y}_i(t+j) + b_k(x(t)), \; k \ge 1. \tag{6.14}
$$

Proof. Sufficiency. Apply the inversion algorithm to S. From (6.14) we obtain at the αth step of the algorithm

$$
\begin{aligned}
\tilde{y}_1(t+1) &= \tilde{a}_1^*(x(t), u(t)) \\
&\cdots \\
\tilde{y}_\alpha(t+\alpha) &= \sum_{i=1}^{\alpha-1} \sum_{j=i+1}^{\alpha} K_{ij}^\alpha \tilde{y}_i(t+j) + \tilde{a}_\alpha^*(x(t), u(t)).
\end{aligned}
\tag{6.15}
$$

Note that by the Inversion Algorithm, possibly after reordering the inputs, the Jacobian matrix of the right hand side of (6.15) with respect to $u^1 = (u_1, \ldots, u_{\rho^*})^T$ around the regular equilibrium point (x^0, u^0) has full row rank $\rho_\alpha = \rho^*$. Moreover,

$$\tilde{a}_k^*(x^0,u^0) = \tilde{y}_k^0 - \sum_{i=1}^{k-1}\left(\sum_{j=i+1}^{k} K_{ij}^k\right)\tilde{y}_i^0 := v_k^{10}\,,\ k=1,\ldots,\alpha\,.$$

By Implicit Function Theorem, the following system of equations

$$\begin{aligned} v^1(t) &= \begin{bmatrix} \tilde{a}_1^*(x(t),u(t)) \\ \ldots \\ \tilde{a}_\alpha^*(x(t),u(t)) \end{bmatrix} \\ v^2(t) &= u^2(t) \end{aligned} \tag{6.16}$$

can be solved for $u^1(t)$ uniquely around the point (x^0,u^0,v^0), where $v^0 = [v^{10,T},v^{20,T}]^T$, $v^{10}=[v_1^{10,T}\ldots v_\alpha^{10,T}]$ and $v^{20}=u^{20}$. The solution of (6.16)

$$\begin{aligned} u^1(t) &= \varphi(x(t),v^1(t),u^2(t)) \\ u^2(t) &= v^2(t) \end{aligned} \tag{6.17}$$

which is such that

$$v^1(t) \equiv \begin{bmatrix} \tilde{a}_1^*(x(t),\varphi(x(t),v^1(t),u^2(t)),u^2(t)) \\ \ldots \\ \tilde{a}_\alpha^*(x(t),\varphi(x(t),v^1(t),u^2(t)),u^2(t)) \end{bmatrix}.$$

and $\partial\varphi(\cdot)/\partial v^1$ is nonsingular.

Notice that $\varphi: M_1 \to M_2$ is defined for some (possible small) neighbourhoods M_1 and M_2 of (x^0,v^{10},u^{20}) and of u^{10}.

Now, it is easy to see that the feedback (6.17) applied to (3.1) yields locally around (x^0,u^0,y^0) for $i=1,\ldots,\alpha$

$$\begin{gathered} \tilde{y}_1(t+1) = v_1^1(t) \\ \ldots \\ \tilde{y}_\alpha(t+\alpha) - \sum_{i=1}^{\alpha-1}\sum_{j=i+1}^{\alpha} K_{ij}^\alpha \tilde{y}_i(t+j) = v_\alpha^1(t)\,,\quad 0 \le t \le t_F\,. \end{gathered}$$

Moreover, inspection of the inversion algorithm gives that for the compensated system (3.1), (6.17) we have that

$$\hat{y}_k(t+k) = \sum_{i=1}^{k}\sum_{j=i}^{k} K_{ij}^k \tilde{y}_i(t+j) + b_k(x(t))\,,\ k \ge 0\,. \tag{6.18}$$

Necessity. This is clear. ∎

Example 6.8 Consider the system

$$\begin{aligned} x_1(t+1) &= x_2(t)u_1(t) + x_3(t) \\ x_2(t+1) &= 2u_1(t)x_2(t) \\ x_3(t+1) &= x_4(t) \\ x_4(t+1) &= x_4(t)u_2(t) + u_1(t) \\ y_1(t) &= x_1(t)\,,\ y_2(t) = x_2(t)\,. \end{aligned}$$

Applying the inversion algorithm to this system, we obtain

$$\begin{aligned} y_1(t+1) &= x_2(t)u_1(t) + x_3(t) \\ y_2(t+1) &= 2y_1(t+1) - 2x_3(t) \\ y_2(t+2) &= 2y_1(t+2) - 2x_4(t) \\ y_2(t+3) &= 2y_1(t+3) - 2x_4(t)u_2(t) - 2u_1(t) . \end{aligned}$$

Note that for this system the conditions (6.14) of Theorem 6.7 are satisfied and a desired static state feedback can be obtained by solving the following system of equations (see (6.16))

$$\begin{aligned} v_1(t) &= x_2(t)u_1(t) + x_3(t) \\ v_2(t) &= -2x_4(t)u_2(t) - 2u_1(t) \end{aligned}$$

with respect to $u_1(t)$ and $u_2(t)$. Solving the above, we obtain the feedback

$$u_1(t) = \frac{v_1(t) - x_3(t)}{x_2(t)}, \quad u_2(t) = \frac{-2v_1(t) + 2x_3(t) - v_2(t)x_2(t)}{2x_4(t)x_2(t)},$$

which yields

$$\begin{aligned} y_1(t+1) &= v_1(t) \\ y_2(t+1) &= 2v_1(t) - 2x_3(t), \quad y_2(t+2) = 2v_1(t+1) - 2x_4(t), \\ y_2(t+3) &= 2v_1(t+2) + v_2(t) . \end{aligned}$$

6.3 The Singh Compensator

The discrete-time equivalent of the so-called Singh compensator can be obtained as follows. Consider the nonlinear system (5.1) and let (x^0, u^0) be a strongly regular equilibrium point of (5.1) with respect to the Inversion Algorithm. Apply the Inversion Algorithm to (5.1). This yields at the last step:

$$\begin{bmatrix} \tilde{y}_1(t+1) \\ \tilde{y}_2(t+2) \\ \vdots \\ \tilde{y}_\alpha(t+\alpha) \end{bmatrix} = A_\alpha(x(t), u(t), \{\tilde{y}_i(t+j),\ 1 \le i \le \alpha-1,\ i+1 \le j \le \alpha\}), \quad (6.19)$$

and

$$\hat{y}_\alpha(t+\alpha) = \psi_\alpha(x(t), \{\tilde{y}_i(t+j),\ 1 \le i \le \alpha,\ i \le j \le \alpha\}), \quad (6.20)$$

where

$$A_\alpha(x(t), u(t), \{\tilde{y}_i(t+j),\ 1 \le i \le \alpha-1,\ i+1 \le j \le \alpha\}) = \\ = \begin{bmatrix} \tilde{a}_1(x(t), u(t)) \\ \tilde{a}_2(x(t), u(t), \tilde{y}_1(t+2)) \\ \vdots \\ \tilde{a}_\alpha(x(t), u(t), \{\tilde{y}_i(t+j),\ 1 \le i \le \alpha-1,\ i+1 \le j \le \alpha\}) \end{bmatrix}$$

and the Jacobian matrix of the right hand side of (6.19) with respect to u, i.e. $\partial A_\alpha(\cdot)/\partial u$ has full row rank ρ^* on a neighbourhood of the equilibrium point (x^0, u^0). After a possible permutation of the controls we may assume that the matrix $\partial A_\alpha(\cdot)/\partial u^1$ consisting of the first ρ^* columns of $\partial A_\alpha(\cdot)/\partial u$ is invertible on this neighbourhood. Therefore, equation (6.19) can be solved for $u^1(t) = (u_1(t), \ldots, u_{\rho^*}(t))^T$ uniquely around the point (x^0, u^0) by applying the Implicit Function Theorem. Denoting $u^2(t) = (u_{\rho^*+1}(t), \ldots, u_m(t))^T$, we obtain from (6.19)

$$u^1(t) = \varphi(x(t), \{\tilde{y}_i(t+j),\ 1 \le i \le \alpha,\ i \le j \le \alpha\}, u^2(t)) \tag{6.21}$$

which is such that

$$\begin{bmatrix} \tilde{y}_1(t+1) \\ \tilde{y}_2(t+2) \\ \vdots \\ \tilde{y}_\alpha(t+\alpha) \end{bmatrix} \equiv \begin{array}{l} A_\alpha(x(t), \varphi(x(t), \{\tilde{y}_i(t+j),\ 1 \le i \le \alpha,\ i \le j \le \alpha\}, u^2(t)) \\ \{\tilde{y}_i(t+j),\ 1 \le i \le \alpha - 1,\ i+1 \le j \le \alpha\}). \end{array}$$

In order to introduce the Singh compensator, we need the following notation. For $i = 1, \ldots, \rho^*$, let $t + \gamma_i$ be the lowest time instant and $t + \varepsilon_i$ be the greatest time instant in which the ith scalar component y_i of the output y appears in equations (6.21). In this notation we can rewrite (6.21) as follows

$$u^1(t) = \varphi(x(t), \{y_i(t+j),\ 1 \le i \le \rho^*,\ \gamma_i \le j \le \varepsilon_i\}, u^2(t)). \tag{6.22}$$

Now construct the compensator for (5.1) in the following way. Let $x_i^C = (x_{i1}^C, \ldots, x_{i,\varepsilon_i-\gamma_i}^C)^T$, $i = 1, \ldots, \rho^*$, be a vector of dimension $\varepsilon_i - \gamma_i$, v^2 a vector of dimension $m - \rho^*$ and consider the system

$$\begin{aligned} x_{i1}^C(t+1) &= x_{i2}^C(t) \\ &\cdots \\ x_{i,\varepsilon_i-\gamma_i-1}^C(t+1) &= x_{i,\varepsilon_i-\gamma_i}^C(t),\ i = 1, \ldots, \rho^* \\ x_{i,\varepsilon_i-\gamma_i}^C(t+1) &= v_i(t) \\ u^1(t) &= \varphi(x(t), \{x_{ij}^C(t),\ 1 \le j \le \varepsilon_i - \gamma_i, v_i(t),\ 1 \le i \le \rho^*\}, v^2(t)) \\ u^2(t) &= v^2(t). \end{aligned} \tag{6.23}$$

Moreover, in accordance with (6.23) and (6.22) define

$$\begin{aligned} x_{ij}^{C0} &= y_i^0,\ i = 1, \ldots, \rho^*,\ j = 1, \ldots, \varepsilon_i - \gamma_i \\ v_i^0 &= y_i^0,\ i = 1, \ldots, \rho^*, \\ v^{20} &= u^{20}. \end{aligned}$$

The dynamic system (6.23) with the controls $v^1(t) = (v_1, \ldots, v_{\rho^*})$ and v^2, the outputs u^1 and u^2 is called a Singh compensator for (5.1).

Lemma 6.9 *Any Singh compensator defined around a strongly regular equilibrium point (x^0, u^0) of (5.1) is a regular dynamic state feedback for (5.1).*

Proof. Obviously, (6.23) is regular if and only if the system with the input $v^1(t) = (v_1(t), \ldots, v_{\rho^*}(t))^T$ and the output $u^1(t)$

$$\begin{aligned} x_i^C(t+1) &= A_i x_i^C(t) + B_i v_i(t)\,,\ i = 1, \ldots, \rho^*\,, \\ u^1(t) &= \varphi(x(t), x_1^C(t), \ldots, x_{\rho^*}^C(t), v^1(t), v^{20}) \end{aligned} \tag{6.24}$$

where

$$A_i = \begin{bmatrix} 0 & 1 & 0 & \ldots & 0 \\ 0 & 0 & 1 & \ldots & 0 \\ \ldots & \ldots & \ldots & \ldots & \ldots \\ 0 & 0 & 0 & \ldots & 1 \\ 0 & 0 & 0 & \ldots & 0 \end{bmatrix}, \quad B_i = \begin{bmatrix} 0 \\ \vdots \\ 0 \\ 1 \end{bmatrix},$$

defines a (x, x^C)-dependent one-to-one map between variables $v^1(t)$ and $u^1(t)$.

Recall that we obtained the function φ by applying the Implicit Function Theorem. By the same theorem, φ is one-valued function. It remains to show that the inverse mapping of φ is also one-valued. From (6.24) we obtain

$$v_i(t) = x_{i,\varepsilon_i - \gamma_i}^C(t+1) = \ldots = x_{i1}^C(t + \varepsilon_i - \gamma_i)\,. \tag{6.25}$$

Denote the ith row of $\tilde{a}_k$ in the inversion algorithm by a_{ki}. Taking into account that the output equation of (6.24) has been obtained as a solution of (6.19) with $u^2 = u^{20}$, $y_i(t + \gamma_i + j - 1) = x_{ij}^C(t)$, $j = 1, \ldots, \varepsilon_i - \gamma_i$ and $y_i(t + \varepsilon_i) = v_i(t)$, we obtain

$$x_{i1}^C(t) = a_{1i}(x(t), u^1(t), u^{20})\,,\ 1 \le i \le \rho_1\,,$$

and taking into account (6.25),

$$v_i(t) = a_{1i}(x(t + \varepsilon_i - \gamma_i), u^1(t + \varepsilon_i - \gamma_i), u^{20})\,,\ i = 1, \ldots, \rho_1\,. \tag{6.26}$$

Let us denote

$$\begin{aligned} \hat{x}_1^C(t) &= \{x_{i2}^C(t) : \varepsilon_i > 2,\ i = 1, \ldots, \rho_1\} \\ \hat{v}_1(t) &= \{v_i(t) : \varepsilon_i = 2,\ i = 1, \ldots, \rho_1\}\,. \end{aligned}$$

Then we obtain

$$x_{i1}^C(t) = a_{2i}(x(t), u^1(t), u^{20}), \hat{x}_1^C(t), \hat{v}_1(t))\,,\ \rho_1 + 1 \le i \le \rho_2\,,$$

and taking into account (6.25), (6.26),

$$v_i(t) = a_{2i}(x(t + \varepsilon_i - \gamma_i), u^1(t + \varepsilon_i - \gamma_i), u^{20}), \hat{x}_1^C(t + \varepsilon_i - \gamma_i)\,, \hat{v}_1(t + \varepsilon_i - \gamma_i))\,,$$
$$\rho_1 + 1 \le i \le \rho_2\,.$$

Applying the above arguments repeatedly, we prove the theorem. ∎

6.4 Input-Output Decoupling of Discrete-time Nonlinear Systems by Dynamic State Feedback

The system is said to be input-output decoupled if each scalar control variable (input) affects one and only one scalar output variable. If the given system does not possess such a property, then one may try to compensate the original system in order to achieve a decoupled system. Depending on the permitted control laws, either static or dynamic decoupling problems can be formulated. In Section 3.4 we considered static decoupling problem. Our interest here is in the dynamic input-output decoupling problem in which one achieves decoupling by dynamic state feedback.

The local dynamic input-output decoupling problem can be stated as follows.

Definition 6.10 Local dynamic input-output decoupling problem. *Given the system S, described by equations (5.1) around an equilibrium point (x^0, u^0), find if possible, a compensator C defined by equations of the form (1.14) together with an initial state x_0^C and neighbourhoods $O_1 = X^{C0} \times X^0 \times V^0$ of (x^{C0}, x^0, v^0) in $X^C \times X \times V$ and O_2 of u^0 in U, being domain and range of C, such that the closed-loop system $S \circ C$, described by (5.1), (1.14) and initialized at (x_0^C, x_0) is input-output decoupled on $O_1 \times O_2$. That is, the first p components of the new control $v, v_1, \ldots, v_p$, influence independently the p outputs $y_1, \ldots, y_p$ and all other components $v_{p+1}, \ldots, v_m$ affect none of the outputs.*

In this section we shall show that for locally right invertible system the Singh compensator (6.23) solves the input-output decoupling problem.

Theorem 6.11 *Consider the system S described by equations (5.1) around a strongly regular equilibrium point (x^0, u^0) with respect to the inversion algorithm. The dynamic input-output decoupling problem for S is around (x^0, u^0) locally solvable if and only if the system S is locally around (x^0, u^0) FTS-right invertible, or equivalently, $\rho^*(S) = p$.*

Proof. Sufficiency. Suppose that the system (5.1) is locally FTS-right invertible, or equivalently, that $\rho^* = p$.

It is easy to see that the Singh compensator (6.23) with $\rho^* = p$ and with arbitrary initial state, applied to (5.1) yields locally around the equilibrium point for $i = 1, \ldots, p$

$$\begin{aligned} y_i(\gamma_i + j - 1) &= x_{ij}^C(0)\,, \qquad j = 1, \ldots, \varepsilon_i - \gamma_i \\ y_i(t + \varepsilon_i) &= v_i(t) \qquad\quad 0 \le t \le t_F\,. \end{aligned} \tag{6.27}$$

Moreover, the inspection of the inversion algorithm gives that for the compensated system (5.1), (6.23) we have that $y_i(0), \ldots, y_i(\gamma_i - 1)$, $i = 1, \ldots, p$ depend only on $x(0), x^C(0)$ and are therefore independent of the new control. Hence any compensator (6.23) obtained via the inversion algorithm, solves the input-output decoupling problem locally around a strongly regular equilibrium point (x^0, u^0).

Necessity. Assume that there exists a dynamic compensator of the form (1.14) which achieves local input-output decoupling of the closed-loop system (5.1), (1.14) around an equilibrium point (x^0, x^{C0}, v^0). This means that the decoupling matrix of the closed-loop system has full row rank in the neighbourhood of (x^0, x^{C0}, v^0), or equivalently, that the closed-loop system is $(\tilde{d}_1, \ldots, \tilde{d}_p)$-FTS right invertible. The latter implies the local reproducibility property. Given an arbitrary set of sequences $\{y_i^*(t),\ 0 \le t \le t_F\} \subset \mathcal{Y}_i^0$, $i = 1, \ldots, p$, one is able to find the control sequence $\{v^*(t),\ 0 \le t \le t_F\}$ such that the closed-loop system (5.1), (1.14) feeded with these controls produces as output

$$y_i(t) = y_i^*(t)\,, \quad \tilde{d}_i \le t \le t_F\,, \quad i = 1, \ldots, p\,. \tag{6.28}$$

Therefore the original system possesses the same reproducibility property, i.e. there exist controls $\{u^*(t),\ 0 \le t \le t_F\}$ such that (6.28) holds when these controls are applied. This follows from computing the controls $u^*(t)$ from (1.14) with inputs $v^*(t)$. The aforementioned reproducibility property of the original system means that the original system is right invertible, or equivalently, $\rho^* = p$. ∎

6.5 The Solution of the Input-Output Linearization Problem via Dynamic State Feedback

In this section, like in Section 6.2, we study the problem of linearizing the input-output map of a nonlinear system. But unlike Section 6.2 now we try to solve the problem via regular *dynamic* state feedback and adopt a bit more general local viewpoint than in the previous sections. Instead of working around an equilibrium point we work around a certain set of time sequences $\{\bar{x}(t), \bar{u}(t), \bar{y}(t);\ 0 \le t \le t_F\}$ that satisfies (5.1), and where t_F is some given final time instant. This set of time sequences will be called the (reference) trajectory of the system (5.1). Then, using the control sequence $\{u(t);\ 0 \le t \le t_F\}$ with each $u(t)$ sufficiently close to $\bar{u}(t)$, and provided that the initial state x_0 is sufficiently close to $\bar{x}(0)$, we can assure that the states $x(t)$ are sufficiently close to $\bar{x}(t)$ and the outputs $y(t)$ are sufficiently close to $\bar{y}(t)$, both for $0 \le t \le t_F$.

It is obvious from the results of the last section that if the system is invertible then the decoupling compensator solves the input-output linearization problem. However, in this section we do not assume the invertibility of the system and give necessary and sufficient conditions for solvability of the problem around a regular reference trajectory.

The input-output map of the system (5.1) around the given trajectory $\{\bar{x}(t), \bar{u}(t), \bar{y}(t);\ 0 \le t \le t_F\}$ takes the form of a Volterra series expansion

$$y(t) = w^0(t, x_0) + \sum_{j \ge 1} \sum_{i_1, i_2, \ldots, i_j = 1}^{m} \sum_{\tau_1 = 0}^{t-1} \sum_{\tau_2 = 0}^{\tau_1} \cdots$$

$$\cdots \sum_{\tau_j = 0}^{\tau_{j-1}} w^j_{i_1 i_2 \ldots i_j}(t, \tau_1, \ldots, \tau_j, x_0)[u_{i_1}(\tau_1) - \bar{u}_{i_1}(\tau_1)] \ldots [u_{i_j}(\tau_j) - \bar{u}_{i_j}(\tau_j)]\,,$$

where $w^j_{i_1 i_2 \ldots i_j}$ is the jth triangular Volterra kernel.

This system (5.1) is said to have a linear input-output map if its Volterra series expansion reduces to one of the form

$$y(t) = w^0(t, x_0) + \sum_{i=1}^{m} \sum_{\tau=0}^{t-1} w_i^1(t-\tau)[u_i(\tau) - \bar{u}_i(\tau)]\,, 0 \le t \le t_F\,, \tag{6.29}$$

that is, if the input-dependent part of this expansion is linear in u and independent of the initial state x_0. Note that the term $w^0(t, x_0)$ in (6.29) which represents the response to the input $u(t) = \bar{u}(t)$, $0 \le t \le t_F$, need not necessarily be the same as the one of a linear system. However, this term affects the response to different inputs in the same way.

If the system (5.1) is such that (6.29) is not true, we may try to satisfy this property via feedback; that is to find a state feedback compensator such that for every $0 \le t \le t_F$ the input-dependent part of the input-output map of the closed-loop system is linear.

We are looking for an analytic compensator C (dynamic state feedback) with a μ-dimensional state x^C, a new m-dimensional control v, described by equations of the form

$$\begin{aligned} x^C(t+1) &= \xi(x^C(t), x(t), v(t))\,,\; x^C(0) = x_0 \\ u(t) &= \varphi(x^C(t), x(t), v(t)) \end{aligned} \tag{6.30}$$

defined locally around a set (to be found) of time sequences $\{\bar{x}^C(t), \bar{x}(t), \bar{v}(t), \bar{u}(t);$ $0 \le t \le t_F\}$ that satisfies equations (6.30) and correspond in a certain way to the trajectory $\{\bar{x}(t), \bar{u}(t), \bar{y}(t);\; 0 \le t \le t_F\}$. We should like to stress that we allow the expression of the function φ to change from point to point of the trajectory. Similarly, we allow the domain $X^C(t) \times X(t) \times V(t)$ and the range $U(t)$ of the compensator C to be different for different time instances and to depend on the chosen trajectory.

We call the compensator C described by (6.30) regular, if it defines around the set of time sequences $\{\bar{x}(t), \bar{x}^C(t), \bar{v}(t), \bar{u}(t);\; 0 \le t \le t_F\}$ the (x, x^C)-dependent one-to-one map between the variables $v(t)$ and $u(t)$.

The closed-loop system (5.1), (6.30), initialized at (x_0, x_0^C), that is the system

$$\begin{aligned} x(t+1) &= f(x(t), h^C(x^C(t), x(t), v(t)))\,,\; x(0) = x_0, x^C(0) = x_0^C\,, \\ x^C(t+1) &= f^C(x^C(t), x(t), v(t)) \\ y(t) &= h(x(t))\,, \end{aligned} \tag{6.31}$$

is denoted by $S \circ C$.

The closed-loop system (6.31) is said to have a linear input-output map if its Volterra series expansion reduces to one of the form

$$y(t) = w^0(t, x_0, x_0^C) + \sum_{i=1}^{m} \sum_{\tau=0}^{t-1} w_i^1(t-\tau)[v_i(\tau) - \bar{v}_i(\tau)]\,,\; 0 \le t \le t_F\,. \tag{6.32}$$

Definition 6.12 Local dynamic input-output linearization problem. *Given the system (5.1) around the set of time sequences* $\{\bar{x}(t), \bar{u}(t), \bar{y}(t);\; 0 \le t \le t_F\}$ *find, if possible, a regular compensator* C *defined by equations of the form (6.30) together with an initial state* x_0^C*, the set of time sequences* $\{\bar{x}^C(t), \bar{v}(t);\; 0 \le t \le t_F\}$*, and* *neighbourhoods* $\Lambda(t) = X^C(t) \times X(t) \times V(t)$ *of* $(\bar{x}^C(t), \bar{x}(t), \bar{v}(t))$ *and* $U(t)$ *of* $\bar{u}(t)$, *being the domain and the range of* C *for each time instant* t, *so that*

(i) *the outputs of the closed-loop system*

$$y^{S\circ C}(t, x_0, x_0^C, v(0), \ldots, v(t-1)) = w^0(t, x_0, x_0^C) + \\ + \sum_{i=1}^{m} \sum_{\tau=0}^{t-1} w_i^1(t-\tau)[v_i(\tau) - \bar{v}_i(\tau)] \,, 0 \le t \le t_F$$

for every $x_0 \in X(0)$, *and* $v(k) \in V(k)$, $0 \le k \le t-1$;

(ii) *the dimension* μ *of the compensator* C *does not continue to increase as the length* t_F *of the time interval increases, i.e.* $\lim_{t_F \to \infty} \mu \neq \infty$.

Remark. If we allow the compensators C whose dimension μ continues to increase as the length t_F of the time interval increases, a peculiar phenomenon appears. This phenomenon will be described below in the Example 6.19. Namely, if we are working on finite time interval, we can always formally solve the problem without any additional conditions via a compensator with enormously high order. This situation should be, of course, avoided. However, not allowing the dependence of μ on t_F altogether, would exclude the cases when nonlinear evolution may die out in a finite number of steps. (See Example 6.20 below). This situation should be also avoided.

Instrumental in the problem solution is the inversion (structure) algorithm for discrete-time nonlinear system. Firstly, the solvability conditions are expressed in terms of the inversion algorithm. Secondly, the proof of the existence and construction of the dynamic state feedback compensator relies on this algorithm.

An inversion algorithm for discrete-time nonlinear system around an equilibrium point was given in Section 5.1. The difficulties that can arise in performing the algorithm, not around an equilibrium point but around the given trajectory $\{\bar{x}(t), \bar{u}(t), \bar{y}(t);\ 0 \le t \le t_F\}$, are the following.

(i) Certain matrices in the algorithm may have different ranks at different points of the trajectory, or equivalently, the invertibility indices (see Section 5.2 for details about the notion of invertibility indices) may change from point to point of the trajectory.

(ii) Independent rows of certain matrices in the algorithm are not the same for all points of trajectory.

Of course, if we impose the conditions which ensure that the invertibility indices remain constant and independent rows of certain matrices are the same along the given trajectory, then the generalization of the basic algorithm is quite straightforward, and we shall not repeat it here.

Like in the basic inversion algorithm certain constant rank conditions have been imposed to ensure that the algorithm can be applied around a given trajectory. We shall summarize these conditions in the definition of regularity of the trajectory.

Definition 6.13 *We call the trajectory* $\{\bar{x}(t), \bar{u}(t), \bar{y}(t);\ 0 \le t \le t_F\}$ *of the system (5.1) regular with respect to the inversion algorithm if for some specific application of the inversion algorithm*

$$\rho_k = \text{rank}\, \frac{\partial}{\partial u} A_k(\cdot)$$

is constant in some neighbourhood of $\{\bar{x}(t), \bar{u}(t), \bar{y}(t);\ 0 \le t \le t_F\}$.

Like in Lemma 5.4 it can be shown that around a regular trajectory the inversion algorithm terminates in, at most, n steps.

Define the rank ρ^* of the system as $\rho^* = \max\{\rho_k,\ k \ge 1\}$ and let α be defined as the smallest $k \in N$ such that $\rho_k = \rho^*$.

In order to apply the inversion algorithm around the given trajectory $\{\bar{x}(t), \bar{u}(t), \bar{y}(t);\ 0 \le t \le t_F\}$ we need to require that $t_F \ge \alpha$.

The solution will be formulated in terms of the inversion algorithm.

Decompose $\hat{y}_k$ and ψ_k according to

$$\hat{y}_k(k) = \begin{bmatrix} \hat{y}_{k1}(k) \\ \hat{y}_\alpha(k) \end{bmatrix}, \quad \psi_k = \begin{bmatrix} \psi_{k1} \\ \psi_{k2} \end{bmatrix}$$

where

$$\hat{y}_{k1}(k) = \psi_{k1}(x_0, \{y_i(j),\ 1 \le i \le k,\ \gamma_i \le j \le \min(k, \varepsilon_i)\}),$$
$$\hat{y}_\alpha(k) = [y_{\rho^*+1}(k), \ldots, y_p(k)]^T = \psi_{k2}(x_0, \{y_i(j),\ 1 \le i \le k,\ \gamma_i \le j \le k\}).$$

Note that by the inversion algorithm for $k \ge \alpha$, $\hat{y}_k(k) = \hat{y}_\alpha(k)$, $\psi_k = \psi_{k2}$.

Now we are ready to formulate one of our results.

Theorem 6.14 *Consider the system S described by (5.1) around the regular trajectory* $\{\bar{x}(t), \bar{u}(t), \bar{y}(t);\ 0 \le t \le t_F\}$ *with respect to some specific application of the inversion algorithm. The dynamic input-output linearization problem for S is around the regular trajectory locally solvable via the Singh compensator (6.23) if and only if, in case of this specific application of the inversion algorithm for all integers* $0 \le k \le t_F$, $\hat{y}_\alpha(k) = [y_{\rho^*+1}(k), \ldots, y_p(k)]^T$, *obtained at the kth step of the inversion algorithm, i.e.*

$$\hat{y}_\alpha(k) = \psi_{k2}(x_0, \{y_i(j),\ 1 \le i \le \rho_k,\ \gamma_i \le j \le k\}),$$

reduces to the following form

$$\hat{y}_\alpha(k) = \psi_k^0(x_0, \{y_i(j),\ 1 \le i \le \rho_k,\ \gamma_i \le j \le \varepsilon_i - 1\}) + \sum_{i=1}^{\rho_k} \sum_{j=\varepsilon_i}^{k} a_{ij}^k y_i(j), \qquad (6.33)$$

where the elements of $(p - \rho^*)$*-dimensional vectors* a_{ij}^k *are constants.*

Proof. Sufficiency. It is easy to see that the Singh compensator (6.23)[1], with arbitrary initial state, applied to (5.1), yields locally around $\{\bar{x}(t), \bar{u}(t), \{\bar{y}_i(t+j),\ 1 \le i \le \rho^*,\ \gamma_i \le j \le \varepsilon_i\})$ for $i = 1, \ldots, \rho^*$

[1]Note that now in order to obtain the Singh compensator we solve equation (6.19) around the point $(\bar{x}(t), \bar{u}(t), \{\bar{y}_i(t+j),\ 1 \le i \le \rho^*,\ \gamma_i + 1 \le j \le \varepsilon_i\})$ and the function φ is defined around the same point. Moreover, $\bar{x}_{ij}^C(t) = \bar{y}_i(t+\gamma_i+j-1)$, $j = 1, \ldots, \varepsilon_i - \gamma_i$, $i = 1, \ldots, \rho^*$, $\bar{v}_i(t) = \bar{y}_i(t+\varepsilon_i)$.

$$\begin{aligned} y_i(\gamma_i + j - 1) &= x_{ij}^C(0)\,,\ j = 1, \ldots, \varepsilon_i - \gamma_i\,, \\ y_i(t + \varepsilon_i) &= v_i(t)\,,\ 0 \le t \le t_F - \varepsilon_i\,. \end{aligned} \tag{6.34}$$

Moreover, inspection of the inversion algorithm gives that for the compensated system (5.1), (6.23) we have that $y_i(0), \ldots, y_i(\gamma_i - 1)$, $i = 1, \ldots, \rho^*$ only depend on x_0 and x_0^C and are therefore independent of the new controls. Hence, any compensator (6.23) obtained via the inversion algorithm, then linearizes the input-dependent part for the first ρ^* outputs.

Consider now the outputs $(y_{\rho^*+1}, \ldots, y_p)^T = \hat{y}_\alpha = \hat{y}_{\alpha+1} = \hat{y}_{\alpha+2} = \ldots$.

By assumption, $\hat{y}_\alpha(k)$ for $k \ge 0$ reduces to the form

$$\hat{y}_\alpha(k) = \psi_k^0(x_0, \{y_i(j),\ 1 \le i \le \rho_k,\ \gamma_i \le j \le \varepsilon_i - 1\}) + \sum_{i=1}^{\rho_k} \sum_{j=\varepsilon_i}^{k} a_{ij}^k y_i(j)\,,$$

which under the application of the Singh compensator (6.23), yields

$$\hat{y}_\alpha(k) = \psi_k^0(x_0, x_0^C) + \sum_{i=1}^{\rho_k} \sum_{j=\varepsilon_i}^{k} a_{ij}^k v_i(j - \varepsilon_i)\,.$$

So, the input-dependent part of $\hat{y}_\alpha(k)$ for all $k \ge 0$ is linear in v and independent of initial state (x_0, x_0^C). Hence, the Singh compensator also linearizes the input-dependent part of the last $p - \rho^*$ outputs.

Necessity. Let us assume that the Singh compensator defined by (6.23) locally around the regular trajectory of S solves the dynamic input-output linearization problem. Apply the inversion algorithm for S. If we plug the output of (6.23) in (6.19), (6.20), the input-dependent parts of the outputs must be linear in v and independent of the initial extended state (x_0, x_0^C) since the Singh compensator solves the dynamic input-output linearization problem for S. Note that by the inversion algorithm for all $k \ge 0$, $\hat{y}_k(k)$ does not depend on control. More precisely,

$$\begin{aligned} \hat{y}_{k1}(k) &= \psi_{k1}(x_0, \{y_i(j),\ 1 \le i \le \rho_k,\ \gamma_i \le j \le \varepsilon_i - 1\}) \\ \hat{y}_\alpha(k) &= \psi_{k2}(x_0, \{y_i(j),\ 1 \le i \le \rho_k,\ \gamma_i \le j \le k\})\,. \end{aligned}$$

This means that under the application of the Singh compensator

$$\hat{y}_{k1}(k) = \psi_{k1}(x_0, x_0^C)$$

and implies that ψ_{k2} must reduce to the form (6.33). This completes the proof. ∎

In most cases, if $\hat{y}_\alpha(\varepsilon_i)$ depends nonlinearly on $y_i(\varepsilon_i) = v_i(0)$, then $\hat{y}_\alpha(\varepsilon_i + k)$ for all $k \ge 1$ depends nonlinearly on $y_i(\varepsilon_i + k)$. Obviously, in these cases, allowing other compensators does not help to solve the problem of input-output linearization, since by the inversion algorithm the components of $\hat{y}_\alpha$ cannot be affected by control. However, in some exceptional cases (see Example 6.19 below) a following typically discrete-time dead-beat phenomenon may appear. Namely, nonlinear evolution may die out in a finite number of steps. Obviously, the "order of dead-beat" λ is an arbitrary number between zero and the dimension n of the system. In these cases, even if the condition (6.33) is not satisfied, the input-output linearization problem can still be solved provided one allows the other compensators. The next theorem considers the general case.

Theorem 6.15 *Consider the plant S described by (5.1) around the regular trajectory $\{\bar{x}(t), \bar{u}(t), \bar{y}(t);\ 0 \leq t \leq t_F\}$ with respect to some specific application of the inversion algorithm. The dynamic input-output linearization problem for S is locally solvable around the regular trajectory if and only if, in case of this specific application of the inversion algorithm, there exists an integer $0 \leq \lambda \leq t_F - \alpha$ such that for all integers $\alpha + \lambda \leq k \leq t_F$, $\hat{y}_k(k)$, obtained at the kth step of the inversion algorithm, i.e.*

$$\hat{y}_k(k) = \psi_k(x_0, \{y_i(j),\ 1 \leq i \leq \rho^*,\ \gamma_i \leq j \leq k\}),$$

reduces to the following form

$$\hat{y}_k(k) = \psi_k^0(x_0, \{y_i(j),\ 1 \leq i \leq \rho^*,\ \gamma_i \leq j \leq \alpha + \lambda - 1\}) + \sum_{i=1}^{\rho^*} \sum_{j=\alpha+\lambda}^{k} a_{ij}^k y_i(j), \quad (6.35)$$

where the elements of $(p - \rho^*)$-dimensional vectors a_{ij}^k are constants.

Proof. Sufficiency. If the assumption (6.33) of Theorem 6.14 is not satisfied, but the milder assumption (6.35) is, then the Singh compensator does not solve the problem, but there exists a compensator of higher dimension that does the job.

If ψ_k^0, $k \geq 0$, depend nonlinearly on $y_r(\varepsilon_r), \ldots, y_r(\alpha + \lambda - 1)$ for some $r \in$ $\in \{1, \ldots, \rho^*\}$, then we may add $\alpha + \lambda - \varepsilon_r$ time-shifts

$$x^C_{r,\varepsilon_r-\gamma_r}(t+1) = x^C_{r,\varepsilon_r-\gamma_r+1}(t)$$
$$\cdots$$
$$x^C_{r,\alpha-\gamma_r+\lambda-1}(t+1) = x^C_{r,\alpha-\gamma_r+\lambda}(t)$$

into the Singh compensator C, defined by (6.23). Divide the set $\{1, \ldots, \rho^*\}$ into the two parts – I_1 and I_2. Let I_1 contain those values of $i \in \{1, \ldots, \rho^*\}$ in whose case all ψ_k^0, $k \geq 0$, depend linearly on $y_i(\varepsilon_i), \ldots, y_i(\alpha + \lambda - 1)$, and let I_2 contain the remaining values of i. Now, instead of compensator C consider the compensator $C1$ defined by

$$\left.\begin{array}{l}\left.\begin{array}{l}x_{i1}^C(t+1) = x_{i2}^C(t)\\ \dots\\ x_{i,\varepsilon_i-\gamma_i-1}^C(t+1) = x_{i,\varepsilon_i-\gamma_i}^C(t)\,,\\ x_{i,\varepsilon_i-\gamma_i}^C(t+1) = v_i(t)\end{array}\right\} i\in I_1\\ \\ \left.\begin{array}{l}x_{r1}^C(t+1) = x_{r2}^C(t)\\ \dots\\ x_{r,\alpha+\lambda-\gamma_r-1}^C(t+1) = x_{r,\alpha+\lambda-\gamma_r}^C(t)\,,\\ x_{r,\alpha+\lambda-\gamma_r}^C(t+1) = v_r(t)\end{array}\right\} r\in I_2\\ \\ u^1(t) = \varphi_t(x(t),\{x_{ij}^C(t),\ 1\le j\le \varepsilon_i-\gamma_i,\ v_i(t), i\in I_1\},\\ \qquad\qquad \{x_{ij}^C(t),\ 1\le j\le\varepsilon_i-\gamma_i+1,\ i\in I_2\}, v^2(t))\\ \\ u^2(t) = v^2(t)\,.\end{array}\right\} \tag{6.36}$$

It is easy to see that under application of (6.36), ψ_k^0, $k\geq 0$, depends only on x_0 and x_0^C, and therefore, taking also into account (6.35), we may conclude that compensator (6.36) linearizes the input-dependent parts of the last $p-\rho^*$ outputs.

Note that adding time-shifts into the Singh compensator leaves the input-dependent parts of the first ρ^* outputs linear in new control v and independent of the initial state.

Necessity. Let us assume that there exists a dynamic feedback control defined by (6.30) for (5.1) that, locally around the regular trajectory of S, solves the dynamic input-output linearization problem. Apply the inversion algorithm for S. If we plug the output of (6.30) in (6.19), and (6.20), the input-dependent parts of the outputs must be linear in v and independent of initial state since C solves the dynamic input-output linearization problem for S. Note that by the inversion algorithm for all $k\geq 0$,

$$\hat{y}_k(k) = \psi_k(x_0, \{y_i(j),\ 1\le i\le\rho^*,\ \gamma_i\le j\le k\})$$

does not depend on control u. This implies that the input-dependent part of $\hat{y}_k(k)$, $k\geq 0$, must be linear in v and independent of initial extended state (x_0, x_0^C), which by regularity, by the form of the compensator, and especially by (ii) of Definition 6.12, means that there must exist $0\le\lambda\le t_F-\alpha$ such that $\hat{y}_k(k)$, $k\ge\alpha+\lambda$, reduces to the form (6.35). ∎

Corollary 6.16 *For systems that are locally right invertible about a given trajectory (i.e. for systems with $\rho^* = p$) the dynamic input-output linearization problem is always locally solvable.*

Proof. The proof follows from the fact that, in the case of invertible systems, $\psi_k = 0$, for $k\geq\alpha$. ∎

Now, let us compare our necessary and sufficient conditions given in this section with the result of Section 6.2. Comparing (6.14) with (6.35) we see that the conditions (6.35) are less restrictive than (6.14). So, we may conclude that the solvability of the dynamic input-output linearization problem does not necessarily imply that the static input-output linearization problem is solvable. Of course, if the latter problem is solvable, then so is the dynamic input-output linearization problem.

Our procedure for constructing the compensator unfortunately leads us to a dynamic compensator even if static solution exists. That is, the compensator (6.23) does not necessarily reduce to static state feedback, if the static input-output linearization problem is solvable.

The non-minimality of the Singh compensator (6.23) with respect to the input-output linearization problem occurs since it produces a compensated system for which ρ^* outputs are decoupled, i.e. $y_1(t+\varepsilon_1) = v_1(t), \ldots, y_{\rho^*}(t+\varepsilon_{\rho^*}) = v_{\rho^*}(t)$; while this property does not have to hold for a regular feedback that renders input-dependent parts of output components linear. The problem of how to construct the dynamic compensator with minimal dimension that solves the input-output linearization problem remains a topic for future research.

However, we can check condition (6.14) without extra computations during the first part of our procedure, i.e. while applying the inversion algorithm to system (5.1). Then, if condition (6.14) is satisfied, we can, instead of compensator (6.23), construct the static state feedback in correspondence with (6.17). Note, that if condition (6.14) is satisfied, then (6.19) reduces to

$$\begin{aligned}\tilde{y}_1(t+1) &= \tilde{a}_1^*(x(t),u(t))\\ &\ldots\\ \tilde{y}_\alpha(t+\alpha) &= \sum_{i=1}^{\alpha-1}\sum_{j=i+1}^{\alpha} K_{ij}^\alpha \tilde{y}_i(t+j) + \tilde{a}_\alpha^*(x(t),u(t))\,.\end{aligned} \tag{6.37}$$

and a static state feedback compensator $u(t) = \varphi(x(t), v(t))$ can be obtained as a solution of the following system of equations

$$\begin{aligned} v^1(t) &= \begin{bmatrix} \tilde{a}_1^*(x(t),u(t)) \\ \ldots \\ \tilde{a}_\alpha^*(x(t),u(t)) \end{bmatrix}\\ v^2(t) &= u^2(t)\,. \end{aligned} \tag{6.38}$$

Now we present various examples that illustrate our theory.

Example 6.17 Consider the system

$$\begin{aligned} x_1(t+1) &= x_2(t) + u_1(t)\\ x_2(t+1) &= x_3(t)u_1(t)\\ x_3(t+1) &= x_3(t)u_2(t)\\ x_4(t+1) &= x_4(t) + u_1(t)\\ y_1(t) &= x_1(t)\,,\ y_2(t) = x_2(t)\,,\ y_3(t) = x_4(t) \end{aligned}$$

Applying the inversion algorithm to this system, we obtain

$$\begin{aligned} y_1(t+1) &= x_2(t) + u_1(t) \\ y_2(t+2) &= x_3(t)u_2(t)[y_1(t+2) - x_3(t)u_1(t)] \end{aligned}$$

and

$$\begin{aligned} y_3(t+2) &= x_4(t) - x_2(t) - x_3(t)[y_1(t+1) - x_2(t)] + y_1(t+1) + y_1(t+2) \\ y_3(t+k) &= x_4(t) - x_2(t) - x_3(t)[y_1(t+1) - x_2(t)] + \sum_{j=1}^{k} y_1(t+j) - \\ &\quad - \sum_{j=2}^{k-1} y_2(t+j),\ k \geq 3: \end{aligned}$$

The trajectory $\{\bar{x}(t), \bar{u}(t), \bar{y}(t); 0 \leq t \leq t_F\}$ is regular if, for $0 \leq t \leq t_F, \bar{x}_3(t) \neq 0$ and $\bar{x}_3(t)\bar{u}_1(t) \neq \bar{y}_1(t+2)$. Moreover, $\alpha = 2$, $\gamma_1 = 1$, $\varepsilon_1 = 2$, $\gamma_2 = \varepsilon_2 = 2$. Since the conditions of Theorem 6.14 are satisfied, the system is input-output linearizable by the dynamic compensator of the form (6.23) with arbitrary initial state $x^C(0) = x_0^C$ such that $x_0^C \neq x_{20} + v_1(0)/x_{30}$

$$\begin{aligned} x^C(t+1) &= v_1(t) \\ u_1(t) &= x^C(t) - x_2(t) \\ u_2(t) &= \frac{v_2(t)}{x_3(t)\{v_1(t) - x_3(t)[x^C(t) - x_2(t)]\}}\,. \end{aligned} \tag{6.39}$$

Applying this compensator, it is easy to see that the input-dependent parts of the output components are linear in new control v and independent of initial state (x_0, x_0^C):

$$\begin{aligned} y_1(t+1) &= x^C(t)\,,\ y_1(t+2) = v_1(t)\,, \\ y_2(t+1) &= x_3(t)[x^C(t) - x_2(t)]\,,\ y_2(t+2) = v_2(t)\,, \\ y_3(t+1) &= x_4(t) + x^C(t) - x_2(t)\,, \\ y_3(t+2) &= x_4(t) - x_2(t) - x_3(t)[x^C(t) - x_2(t)] + x^C(t) + v_1(t)\,, \\ y_3(t+k) &= x_4(t) - x_2(t) - x_3(t)[x^C(t) - x_2(t)] + x^C(t) + \\ &\quad + \sum_{j=2}^{k} v_1(t+j-2) - \sum_{j=2}^{k-1} v_2(t+j-2)\,,\ k \geq 3\,. \end{aligned}$$

Note that for this system, conditions (6.14) are not satisfied. So, there exist no static state feedback that solves input-output linearization problem.

The Singh compensator (6.23) is certainly not a regular dynamic state feedback of minimal dimension, which solves the input-output linearization problem. This is illustrated by the following example.

Example 6.18 Consider the system

$$\begin{aligned}
x_1(t+1) &= x_3(t)x_4(t) + u_1(t)\\
x_2(t+1) &= x_3(t) + 2u_1(t)\\
x_3(t+1) &= u_2(t)x_3(t)\\
x_4(t+1) &= x_4(t)\\
x_5(t+1) &= u_1(t) + x_6(t)\\
x_6(t+1) &= x_7(t)\\
x_7(t+1) &= u_3(t)\\
y_1(t) = x_1(t)\,,\ & y_2(t) = x_2(t)\,,\ y_3(t) = x_5(t)\,.
\end{aligned}$$

Applying the inversion algorithm to this system, we obtain

$$\begin{aligned}
y_1(t+1) &= x_3(t)x_4(t) + u_1(t)\\
y_2(t+1) &= 2y_1(t+1) + [1 - 2x_4(t)]x_3(t)\\
y_3(t+1) &= y_1(t+1) - x_3(t)x_4(t) + x_6(t)\\
y_2(t+2) &= 2y_1(t+2) + x_3(t)u_2(t)[1 - 2x_4(t)]\\
y_3(t+2) &= y_1(t+2) + x_7(t) - \frac{x_4(t)}{1-2x_4(t)}[y_2(t+2) - 2y_1(t+2)]\\
y_3(t+3) &= y_1(t+3) + u_3(t) - \frac{x_4(t)}{1-2x_4(t)}[y_2(t+3) - 2y_1(t+3)]\,.
\end{aligned}$$

This system has full rank ρ^* equal to three and $\gamma_1 = 1$, $\varepsilon_1 = 3$, $\gamma_2 = 2, \varepsilon_2 = 3$, $\gamma_3 = \varepsilon_3 = 3$. By Theorem 6.14 (see also Corollary 6.16) the system is input-output linearizable by dynamic state feedback of the form (6.23) with dimension $\mu = (\varepsilon_1 - \gamma_1) + (\varepsilon_2 - \gamma_2) + (\varepsilon_3 - \gamma_3) = 3$. However, it can be easily checked that another regular first order dynamic compensator with arbitrary initial state x_0^C

$$\begin{aligned}
x^C(t+1) &= v_2(t)\,,\ x^C(0) = x_0\,,\\
u_1(t) &= v_1(t) - x_3(t)x_4(t)\\
u_2(t) &= \frac{x^C(t)}{x_3(t)[1-2x_4(t)]}\\
u_3(t) &= v_3(t) + \frac{v_2(t)x_4(t)}{1-2x_4(t)}
\end{aligned} \tag{6.40}$$

also solves the input-output linearization problem. Actually, applying (6.40) yields

$$\begin{aligned}
y_1(t+1) &= v_1(t)\\
y_2(t+1) &= x_3(t) + 2v_1(t) - 2x_3(t)x_4(t)\,,\\
y_2(t+2) &= x^C(t) + 2v_1(t+1)\,,\\
y_2(t+3) &= v_2(t) + 2v_1(t+2)\\
y_3(t+1) &= x_6(t) + v_1(t) - x_3(t)x_4(t)\,,\\
y_3(t+2) &= x_7(t) + v_1(t+1) - \frac{x_4(t)x^C(t)}{1-2x_4(t)}\,,\\
y_3(t+3) &= v_3(t) + v_1(t+2)\,.
\end{aligned}$$

Note, that we obtained the compensator (6.40) via solving the following system of equations for $u(t)$

$$\begin{aligned} v_1(t) &= x_3(t)x_4(t) + u_1(t) \\ v_2(t-1) &= x_3(t)[1 - 2x_4(t)]u_2(t) \\ v_3(t) &= u_3(t) - \frac{x_4(t)}{1-2x_4(t)}v_2(t) \end{aligned}$$

and replacing $v_2(t-1)$ by $x^C(t)$.

Allowing compensators C whose dimension μ increases as the length t_F of the time interval increases, a peculiar phenomenon appears. We shall explain this phenomenon via the following example.

Example 6.19 Consider the system

$$\begin{aligned} x_1(t+1) &= x_2(t) + u_1(t) \\ x_2(t+1) &= x_3(t)u_1(t) \\ x_3(t+1) &= u_2(t) \\ x_4(t+1) &= x_5(t)u_1(t) \\ x_5(t+1) &= x_5(t) \\ y_1(t) = x_1(t)\,,\; & y_2(t) = x_2(t)\,,\; y_3(t) = x_4(t)\,. \end{aligned}$$

The Singh compensator C, defined by (6.23)

$$\begin{aligned} x_1^C(t+1) &= v_1(t) \\ u_1(t) &= x_1^C(t) - x_2(t)\,, \\ u_2(t) &= \frac{v_2(t)}{v_1(t) - x_3(t)[x_1^C(t) - x_2(t)]} \end{aligned}$$

does not solve the input-output linearization problem because the input-dependent parts of $y_3(2), y_3(3), y_3(4)$ etc. are not linear:

$$\begin{aligned} y_3(2) &= x_5(0)\{v_1(0) - x_3(0)[x_1^C(0) - x_2(0)]\} \\ y_3(3) &= x_5(0)[v_1(1) - v_2(0)] \\ y_3(4) &= x_5(0)[v_1(2) - v_2(1)]\,. \end{aligned}$$

Now, add a time-shift into the compensator C

$$x_1^C(t+1) = x_2^C(t)$$

and consider the compensator $C1$:

$$\begin{aligned} x_1^C(t+1) &= x_2^C(t) \\ x_2^C(t+1) &= v_1(t) \\ u_1(t) &= x_1^C(t) - x_2(t)\,, \\ u_2(t) &= \frac{v_2(t)}{x_2^C(t) - x_3(t)[x_1^C(t) - x_2(t)]}\,. \end{aligned}$$

With this compensator the input-dependent part of

$$y_3(2) = x_5(0)\{x_2^C(0) - x_3(0)[x_1^C(0) - x_2(0)]\}$$

is equal to zero, and therefore, can be considered linear. Of course

$$y_3(3) = x_5(0)[v_1(0) - v_2(0)]$$

is still not linear. But, by adding two more time-shifts into the compensator, i.e. using $C2$ defined by the equations

$$\begin{aligned}
x_1^C(t+1) &= x_2^C(t), \\
x_2^C(t+1) &= x_3^C(t), \\
x_3^C(t+1) &= v_1(t), \\
x_4^C(t+1) &= v_2(t), \\
u_1(t) &= x_1^C(t) - x_2(t), \\
u_2(t) &= \frac{x_4^C(t)}{x_2^C(t) - x_3(t)[x_1^C(t) - x_2(t)]}
\end{aligned}$$

we can make also $y_3(3)$ independent on new control v

$$y_3(3) = x_5(0)[x_3^C(0) - x_4^C(0)].$$

As we are working on a finite time interval, we can always formally solve the input-output linearization problem around the regular trajectory without any additional conditions. Of course, we must use for this purpose a compensator with enormously high dimension.

The next example considers the case when the Singh compensator does not solve the problem, but there exists a compensator of higher dimension that does the job.

Example 6.20 Consider the system

$$\begin{aligned}
x_1(t+1) &= u_1(t) \\
x_2(t+1) &= x_3(t)u_1(t) \\
x_3(t+1) &= u_2(t) \\
x_4(t+1) &= x_5(t)u_1(t) + x_1(t) \\
x_5(t+1) &= x_6(t) \\
x_6(t+1) &= x_7(t) - x_8(t) \\
x_7(t+1) &= x_8(t) \\
x_8(t+1) &= x_8(t) \\
y_1(t) &= x_1(t),\ y_2(t) = x_2(t),\ y_3(t) = x_4(t).
\end{aligned}$$

Applying the inversion algorithm to this system, we obtain

$$\begin{aligned}
y_1(t+1) &= u_1(t) \\
y_2(t+2) &= u_2(t)y_1(t+2)
\end{aligned}$$

and

$$\begin{aligned}
y_2(t+1) &= x_3(t)y_1(t+1)\\
y_3(t+1) &= x_5(t)y_1(t+1) + x_1(t)\\
y_3(t+2) &= x_6(t)y_1(t+2) + y_1(t+1)\\
y_3(t+3) &= [x_7(t) - x_8(t)]y_1(t+3) + y_1(t+2)\\
y_3(t+4) &= y_1(t+3)\text{etc.}
\end{aligned}$$

So, $\alpha = 2, \gamma_1 = 1, \varepsilon_1 = 2, \gamma_2 = \varepsilon_2 = 2$.

Since the condition of Theorem 6.14 is not satisfied, the input-dependent part of the output of the system cannot be linearized by the Singh compensator

$$\begin{aligned}
x_1^C(t+1) &= v_1(t)\\
u_1(t) &= x_1^C(t), \quad u_2(t) = v_2(t)/v_1(t).
\end{aligned} \tag{6.41}$$

However, the following compensator

$$\begin{aligned}
x_1^C(t+1) &= x_2^C(t)\\
x_2^C(t+1) &= x_3^C(t)\\
x_3^C(t+1) &= v_1(t)\\
u_1(t) &= x_1^C(t), \quad u_2(t) = v_2(t)/x_2^C(t)
\end{aligned} \tag{6.42}$$

of higher dimension than the Singh compensator (6.41) solves the problem of linearizing the input-dependent part of the output. Really, using this compensator, we can make also $y_3(t+2)$ and $y_3(t+3)$ independent of new control (so they are no more nonlinear in v):

$$\begin{aligned}
y_3(t+2) &= x_6^C(t)x_2^C(t) + x_1^C(t)\\
y_3(t+3) &= [x_7(t) - x_8(t)]x_3^C(t) + x_2^C(t)\\
y_3(t+4) &= x_3^C(t)\text{etc.}
\end{aligned}$$

6.6 Immersion a Discrete-time Nonlinear System by Regular Dynamic State Feedback into a Linear System

There is another slightly different notion of input-output linearity, expressed via the notion of immersion into a linear system. Suppose that we have a discrete time linear system L

$$\begin{aligned}
x^*(t+1) &= Ax^*(t) + Bv(t), \ x^* \in R^q, x^*(0) = x_0^*,\\
y(t) &= Cx^*(t).
\end{aligned} \tag{6.43}$$

The input-output map of (6.43) has the form

$$y(t) = CA^t x_0^* + \sum_{\tau=0}^{t-1} CA^{t-\tau-1}Bv(\tau).$$

Definition 6.21 *The system $S \circ C$ is said to be locally around the trajectory $\{\bar{x}(t), \bar{x}^C(t), \bar{v}(t), \bar{y}(t);\ 0 \le t \le t_F\}$ immersed into a linear system, if there exist a triplet of matrices (A, B, C), respectively of size $q \times q$, $q \times m$, $p \times q$, and an analytic map $x^* = Q(x, x^C)$ defined around $(\bar{x}(0), \bar{x}^C(0))$ such that*

$$y(t) = CA^tQ(x_0, x_0^C) + \sum_{\tau=0}^{t-1} CA^{t-\tau-1}Bv(\tau)\,,\ 0 \le t \le t_F\,. \tag{6.44}$$

Here Q represents a map from $X \times X^C$ to $X^* \subset R^q$, which is said to be an immersion map.

In other words, the system $S \circ C$ is immersed into a linear system, if there exist a linear system L and an analytic map $x^* = Q(x, x^C)$ such the systems $S \circ C$ and L, initialized at (x_o, x_0^C) and $x_0^* = Q(x_0, x_0^C)$ respectively, have exactly the same input-output map.

Definition 6.22 *The system S is said to be locally around the trajectory $\{\bar{x}(t), \bar{u}(t), \bar{y}(t);\ 0 \le t \le t_F\}$ immersed by regular (analytic) dynamic state feedback into a linear system if there locally exists a regular (analytic) dynamic state feedback C of the form (6.30) defined locally around the trajectory $\{\bar{x}^C(t), \bar{x}(t), \bar{v}(t), \bar{u}(t);\ 0 \le t \le t_F\}$ such that*

(i) the resulting closed-loop system $S \circ C$, described by (6.31), is locally around the trajectory $\{\bar{x}(t), \bar{x}^C(t), \bar{v}(t), \bar{y}(t);\ 0 \le t \le t_F\}$ immersed into a linear system,

(ii) the μ-dimensional compensator C that solves the problem for $t_F = n$, solves it also for each finite $t_F > n$.

The objective of immersion by regular dynamic state feedback into a linear system is not only to make the input-dependent part of the output of closed-loop system linear in the new input v and independent of initial state as in (6.32), but to make output sequence jointly linear in the new input v and some analytic function Q of the initial (extended) state $\mathrm{col}(x_0, x_0^C)$ as in (6.44).

Clearly, immersion by regular dynamic state feedback into a linear system is stronger (requires more) than linearization of the input-dependent part of the output by regular dynamic state feedback. So, a solution to the second problem may exist even in cases where the solution to the first problem does not exist. Necessary and sufficient conditions for the second problem we presented in Section 6.5. The purpose of this section is to find necessary and sufficient conditions to the first problem, that is to the problem of local immersion a discrete-time nonlinear system via a regular dynamic state feedback into a linear system.

Remark. The static immersion problem is obtained when $\mu = \dim x^C = 0$.

Remark. If we allow the compensators C whose dimension μ monotonously increases as the length t_F of the time interval increases, we can always solve the problem posed in last section without any additional conditions. In the problem of immersion by feedback into a linear system, the compensators with high dimension cannot in general be combined with immersion maps of high dimension so that the problem would be always solvable. This will be explained in the Example 6.29 below. However, we included (ii) in Definition 6.21 in order to make our results on immersion into a linear system comparable with the results on linearizing the input-dependent part of the output.

Theorem 6.23 *Consider the system S described by equations (5.1) around the regular trajectory $\{\bar{x}(t), \bar{u}(t), \bar{y}(t);\ 0 \le t \le t_F\}$ with respect to some specific application of the inversion algorithm and suppose that $t_F \ge n$, the dimension of the state of the system. The system S is locally around the regular trajectory immersed by the regular dynamic state feedback into a linear system if and only if in case of this specific application of the inversion algorithm there exists an integer $0 \le \lambda \le n - \alpha$ such that*

(A) for all integers $\alpha + \lambda \le k \le t_F$, $\hat{y}_k(k)$, obtained at the kth step of the inversion algorithm, i.e.

$$\hat{y}_k(k) = \psi_k(x(0), \{y_i(j),\ 1 \le i \le \rho^*,\ \gamma_i \le j \le k\}),$$

reduces to the following form

$$\hat{y}_k(k) = \psi_k^0(x(0), \{y_i(j),\ 1 \le i \le \rho^*,\ \gamma_i \le j \le \alpha + \lambda - 1\}) + \sum_{i=1}^{\rho^*} \sum_{j=\alpha+\lambda}^{k} a_{ij}^k y_i(j),$$

where the elements of $(p - \rho^*)$-dimensional vectors a_{ij}^k are constants and

(B) $\dim \psi$ *is finite, where ψ is the vector space (over R) generated by functions ψ_k^0, $k \ge 0$.*

Proof. Sufficiency. Suppose that (A) and (B) hold.

The condition (A) is actually sufficient condition for local linearizability of the input-dependent part of the output by the regular dynamic state feedback (see Theorem 6.15 in Section 6.5).

Note that if (A) holds then the following dynamic state feedback, obtained on the basis of the Singh compensator by adding time-shifts into it, linearizes the input-dependent part of the output

$$\begin{aligned} x_{i1}^C(t+1) &= x_{i2}^C(t) \\ &\dots \\ x_{i,\delta_i-\gamma_i-1}^C(t+1) &= x_{i,\delta_i-\gamma_i}^C(t),\ i \in \{1, \dots, \rho^*\} \\ x_{i,\delta_i-\gamma_i}^C(t+1) &= x_{i,\delta_i-\gamma_i+1}^C(t) \\ &\dots \\ x_{i,\alpha+\lambda-\gamma_i-1}^C(t+1) &= x_{i,\alpha+\lambda-\gamma_i}^C(t) \end{aligned} \tag{6.45}$$

$$\begin{aligned} x^C_{i,\alpha+\lambda-\gamma_i}(t+1) &= v_i(t) \\ u^1(t) &= \varphi_t(x(t), \{x^C_{ij}(t),\ 1 \le i \le \rho^*,\ 1 \le j \le \alpha+\lambda-\gamma_i\}, v(t)) \\ u^2(t) &= v^2(t)\,. \end{aligned}$$

Note that the compensator of the form (6.45) was chosen in order to make the proof technically more simple and illuminating. Actually a lower order compensator may solve the problem. See Remark after the proof of theorem.

The compensator (6.45) with arbitrary initial state, applied to (5.1), yields locally around the reference trajectory

$$\begin{aligned} y_i(\gamma_i+j-1) &= x^C_{ij}(0),\ i=1,\dots,\rho^*,\ j=1,\dots,\alpha+\lambda-\gamma_i \\ y_i(t+\alpha+\lambda) &= v_i(t),\ 0 \le t \le t_F-\alpha-\lambda\,. \end{aligned} \tag{6.46}$$

By (6.46) we can rewrite the functions ψ_k, $1 \le k \le \alpha+\lambda$, as follows:

$$\begin{aligned} &\psi_k(x(0), \{y_i(j),\ 1 \le i \le \rho_k,\ \gamma_i \le j \le k\}) = \\ &\quad = \psi_k(x(0), x^C(0)),\ k=1,\dots,\alpha+\lambda-1\,, \\ &\psi_{\alpha+\lambda}(x(0), \{y_i(j),\ 1 \le i \le \rho^*,\ \gamma_i \le j \le \alpha+\lambda\}) \overset{\text{by}(A)}{=} \\ &\quad \overset{\text{by}(A)}{=} \psi^0_{\alpha+\lambda}(x(0), x^C(0)) + \sum_{i=1}^{\rho^*} a^{\alpha+\lambda}_{i,\alpha+\lambda} v_i(0)\,. \end{aligned} \tag{6.47}$$

Since $\dim\psi$ is finite by assumption, we may choose a basis

$$\{x^C, \psi^0_0(x), \psi^0_{11}(x,x^C), \psi^0_{21}(x,x^C), \dots, \psi^0_{k1}(x,x^C)\}$$

for

$$\psi^+ = \text{span}_R\{x^C, \psi^0_0(x), \psi^0_1(x,x^C), \dots, \psi^0_k(x,x^C), \dots\}\,.$$

By inversion algorithm, $k \ge \alpha+\lambda-1$ and ψ^0_{r1}, $r=1,\dots,k$ are $(p-\rho_r-\beta_r)$-dimensional subvectors (for some suitable $0 \le \beta_r \le p-\rho_{r+1}$) of $(p-\rho_r)$-vectors ψ^0_r.

Let us define the analytic map $Q: R^{n+\mu} \to R^q : \text{col}(x, x^C) \to x^*$ by

$$Q(x,x^C) = \begin{bmatrix} \psi^0_0(x) \\ \psi^0_{11}(x,x^C) \\ \vdots \\ \psi^0_{k1}(x,x^C) \\ x^C \end{bmatrix} = \begin{bmatrix} x^{*1} \\ x^{*2} \\ \vdots \\ x^{*k+1} \\ x^{*k+2} \end{bmatrix} = x^*\,. \tag{6.48}$$

Next we prove that the analytic map $Q(x,x^C)$ just defined serves as an immersion map. For this purpose, we shall first show that variable x^* satisfies a linear difference equation, and then we show that the input-output map of this linear difference equation is the same as that of the compensated system (5.1), (6.45). In order to see that x^* satisfies a linear difference equation, one has simply to take into account (6.47), condition (B), and to recall the steps of the inversion algorithm (IA). Indeed,

$$x^{*1}(t+1) = \psi_0^0(x(t+1)) = h(f(x(t), \varphi(x^C(t), x(t), v(t)))) =$$

$$\overset{\text{by}IA,(6.46)}{=} \begin{bmatrix} x_{11}^C(t) \\ \vdots \\ x_{\rho_1,1}^C(t) \\ \psi_{11}^0(x(t), x^C(t)) \\ \psi_{12}^0(x(t), x^C(t)) \end{bmatrix} \overset{\text{by}(6.48),(B)}{=} \begin{bmatrix} x_{11}^{*\alpha+\lambda+1}(t) \\ \vdots \\ x_{\rho_1,1}^{*\alpha+\lambda+1}(t) \\ x^{*2}(t) \\ \sum_{j=0}^{k+1} b_j^1 x^{*j+1}(t) \end{bmatrix},$$

where ψ_{12}^0 is a subvector of ψ_1^0, not belonging to the basis of ψ.

$$x^{*2}(t+1) = \psi_1(x(t+1), x^C(t+1)) =$$

$$\overset{\text{by}IA,(6.46)}{=} \begin{bmatrix} x_{\rho_1+1,1}^C(t) \\ \vdots \\ x_{\rho_2,1}^C(t) \\ \psi_{21}^0(x(t), x^C(t)) \\ \psi_{22}^0(x(t), x^C(t)) \end{bmatrix} \overset{\text{by}(6.48),(B)}{=} \begin{bmatrix} x_{\rho_1+1,1}^{*\alpha+\lambda+1}(t) \\ \vdots \\ x_{\rho_2,1}^{*\alpha+\lambda+1}(t) \\ x^{*3}(t) \\ \sum_{j=0}^{k+1} b_j^2 x^{*j+1}(t) \end{bmatrix}, \tag{6.49}$$

where ψ_{22}^0 is a subvector of ψ_2^0, not belonging to the basis of ψ.

....

$$x^{*\alpha+\lambda}(t+1) = \psi_{\alpha+\lambda-1,1}(x(t+1), x^C(t+1)) \overset{\text{by}IA;(6.46),(6.47)}{=}$$
$$= \begin{bmatrix} \psi_{\alpha+\lambda,1}^0(x(t), x^C(t)) \\ \psi_{\alpha+\lambda,2}^0(x(t), x^C(t)) \end{bmatrix} + \sum_{i=1}^{\rho^*} a_{i,\alpha+\lambda}^{\alpha+\lambda,2} v_i(t) \overset{\text{by}(B)}{=}$$
$$= \begin{bmatrix} x^{*\alpha+\lambda+1}(t) \\ \sum_{j=0}^{k+1} b_j^{\alpha+\lambda} x^{*j+1}(t) \end{bmatrix} + \sum_{i=1}^{\rho^*} a_{i,\alpha+\lambda}^{\alpha+\lambda,2} v_i(t).$$

where $\psi_{\alpha+\lambda,2}^0$ is a subvector of ψ_2^0, not belonging to the basis of ψ.

....

$$x^{*k+1}(t+1) = \psi_{k1}^0(x(t+1), x^C(t+1)) \overset{\text{by}(IA),(A)}{=}$$
$$= \psi_{k1}(x(t+1), x^C(t+1)) - \sum_{i=1}^{\rho^*} \sum_{j=\alpha+\lambda}^{k} a_{ij}^{k1} y_i(t+j+1) =$$
$$= \psi_{k+1,2}^0(x(t), x^C(t)) + \sum_{i=1}^{\rho^*} \sum_{j=\alpha+\lambda}^{k+1} a_{ij}^{k+1,2} y_i(t+j) -$$
$$- \sum_{i=1}^{\rho^*} \sum_{j=\alpha+\lambda}^{k} a_{ij}^{k1} y_i(t+j+1) =$$
$$\overset{\text{by}(B),(6.46)}{=} \sum_{j=0}^{k+1} b_j^{k+1} x^{*j+1}(t) + \sum_{i=1}^{\rho^*} a_{i,\alpha+\lambda}^{k+1,2} v_i(t)$$

since by the inversion algorithm

$$a^{k+1,2}_{i,j+1} = a^{k1}_{ij}\,.$$

Note that by definition of $\psi_0(x)$, the output of the closed-loop system (5.1), (6.45) can be expressed as

$$y(t) = x^{*1}(t)\,. \tag{6.50}$$

Now, it is not difficult to see, that the input-output maps of the systems (5.1), (6.45) and (6.49), (6.50), under the initial conditions (x_0, x_0^C) and $Q(x_0, x_0^C) = x_0^*$ respectively, coincide.

Necessity. Suppose that the system (5.1) is locally immersed, by means of regular dynamic state feedback, into a linear system.

Clearly, immersion by feedback into a linear system is stronger than linearization by feedback the input-dependent part of the output. If a solution to the first problem exists, then there also exists a solution to the second problem. The necessity of (A) results from the fact that by Theorem 6.15, (A) is necessary for local solvability of linearization the input-dependent part of the output.

Now, we are going to show the necessity of (B). If the system (5.1) is locally immersed by regular dynamic state feedback into a linear system, then there exists a regular dynamic state feedback C such that the closed-loop system $S \circ C$ is immersed, by means of $Q(x, x^C) : R^{n+\mu} \to R^q$ into a linear system L. At $(x_0, x_0^C) \in R^{n+\mu}$ and $x_0^* = Q(x_0, x_0^C) \in R^q$ the Volterra series of $S \circ C$ given by

$$\begin{aligned} y(t) = w^0(t, x_0, x_0^C) + \sum_{j\geq 1} \sum_{i_1,i_2,\ldots,i_j=1}^{m} \sum_{\tau_1=0}^{t-1} \sum_{\tau_2=0}^{\tau_1} \cdots \\ \cdots \sum_{\tau_j=0}^{\tau_j-1} w^j_{i_1 i_2 \ldots i_j}(t, \tau_1, \ldots, \tau_j, x_0, x_0^C)[v_{i_1}(\tau_1) - \bar{v}_{i_1}(\tau_1)] \ldots [v_{i_j}(\tau_j) - \bar{v}_{i_j}(\tau_j)]\,, \end{aligned} \tag{6.51}$$

and that of L given by

$$y(t) = CA^t x_0^* + \sum_{\tau=0}^{t-1} CA^{t-\tau-1} Bv(\tau)\,,\ t \geq 0$$

coincide by definition. Because $CA^t x_0^*$ corresponds to $w^0(t, x_0, x_0^C) - \sum_{i=1}^{m} \sum_{\tau=0}^{t-1} w_i^1(t, \tau)\bar{v}_i(\tau)$ in (6.51) (or, in (6.44)) and

$$\begin{aligned} \Phi_t(x, x^C) := w^0(t, x, x^C) - \sum_{i=1}^{m} \sum_{\tau=0}^{t-1} w_i^1(t, \tau)\bar{v}_i(\tau) = \\ = \{h \circ \tilde{f}(\cdot, \bar{v}(t-1)) \circ \ldots \circ \tilde{f}(\cdot, \bar{v}(0)) - \sum_{\tau=0}^{t-1} \frac{\partial}{\partial v(\tau)} h \circ \\ \circ \tilde{f}(\cdot, v(t-1)) \circ \ldots \circ \tilde{f}(\cdot, v(0))\big|_{v(k)=\bar{v}(k),\,0\leq k\leq t-1} \bar{v}(\tau)\}(x, x^C)\,, \end{aligned}$$

where $\tilde{f}(x, x^C, v) = \begin{bmatrix} f(x, \varphi(x, x^C, v)) \\ \xi(x^C, x, v) \end{bmatrix}$, the Cayley-Hamilton theorem implies that $\dim \Phi$ is finite, where

$$\Phi = \operatorname{span}_R\{\Phi_0, \Phi_1, \ldots, \Phi_t, \ldots\}.$$

Note that by the inversion algorithm and the dynamic state feedback (6.45) which linearizes the input-dependent part of the output, we have

$$\begin{aligned}
\Phi_0 &= h = \psi_0^0 \\
\Phi_1 &= \operatorname{col}(x_{11}^C, \ldots, x_{\rho_1,1}^C, \psi_1^0(x, x^C)) \\
\Phi_2 &= \operatorname{col}(x_{12}^C, \ldots, x_{\rho_1,2}^C, x_{\rho_1+1,1}^C, \ldots, x_{\rho_2,1}^C, \psi_2^0(x, x^C)) \\
&\ldots \\
\Phi_{\alpha-1} &= \operatorname{col}(x_{1,\alpha-1}^C, \ldots, x_{\rho_1,\alpha-1}^C, x_{\rho_1+1,\alpha-2}^C, \ldots, x_{\rho_2,\alpha-2}^C, \ldots \\
&\quad \ldots, x_{\rho_{\alpha-2}+1,1}^C, \ldots, x_{\rho_{\alpha-1},1}^C, \psi_{\alpha-1}^0(x, x^C)) \\
\Phi_\alpha &= \operatorname{col}(x_{1,\alpha}^C, \ldots, x_{\rho_1,\alpha}^C, x_{\rho_1+1,\alpha-1}^C, \ldots, x_{\rho_2,\alpha-1}^C, \ldots \\
&\quad \ldots, x_{\rho_{\alpha-1}+1,1}^C, \ldots, x_{\rho_\alpha,1}^C, \psi_\alpha^0(x, x^C)), \\
\Phi_{\alpha+1} &= \operatorname{col}(x_{1,\alpha+1}^C, \ldots, x_{\rho_1,\alpha+1}^C, x_{\rho_1+1,\alpha}^C, \ldots, x_{\rho_2,\alpha}^C, \ldots \\
&\quad \ldots, x_{\rho_{\alpha-1}+1,2}^C, \ldots, x_{\rho_\alpha,2}^C, \psi_{\alpha+1}^0(x, x^C)), \\
\Phi_{\alpha+\lambda-1} &= \operatorname{col}(x_{1,\alpha+\lambda-1}^C, \ldots, x_{\rho_1,\alpha+\lambda-1}^C, x_{\rho_1+1,\alpha+\lambda-2}^C, \ldots, x_{\rho_2,\alpha+\lambda-2}^C, \ldots \\
&\quad \ldots, x_{\rho_{\alpha-1}+1,\lambda}^C, \ldots, x_{\rho_\alpha,\lambda}^C, \psi_{\alpha+\lambda-1}^0(x, x^C)), \\
\Phi_{\alpha+\lambda+j} &= \operatorname{col}(0, \ldots, 0, \psi_{\alpha+\lambda+j}^0(x, x^C)), \ j \geq 0.
\end{aligned}$$

The last equalities imply that $\psi \subset \Phi = \psi^+$. Thus, $\dim \psi$ is finite, and the proof is complete. ∎

Remark. Note that the proof of sufficiency is constructive and the compensator C together with the immersion map $Q(x, x^C)$ have been given. However, we should like to stress that the form (6.45) of the compensator C was chosen in order to make the proof technically more simple and illuminating. If we take into account that for every $i \in \{1, \ldots, \rho^*\}$ there may exist different ξ_i, $0 \leq \xi_i \leq \alpha + \lambda - \delta_i$ such that $\hat{y}_k(k)$ takes the form

$$\hat{y}_k(k) = \psi_k^*(x(0), \{y_i(j), 1 \leq i \leq \rho^*, \gamma_i \leq j \leq \delta_i + \xi_i - 1\}) + \sum_{i=1}^{\rho^*} \sum_{j=\delta_i+\xi_i}^{k} a_{ij}^k y_i(j)$$

then we may obtain a lower order compensator than (6.24) and a simpler immersion map than (6.48) that solve the problem. Actually, the compensator C may be chosen as follows

$$
\begin{aligned}
x^C_{i1}(t+1) &= x^C_{i2}(t) \\
&\dots \\
x^C_{i,\delta_i+\xi_i-\gamma_i-1}(t+1) &= x^C_{i,\delta_i+\xi_i-\gamma_i}(t) \\
x^C_{i,\delta_i+\xi_i-\gamma_i}(t+1) &= v_i(t) \\
u^1(t) &= \varphi_t(x(t), \{x^C_{ij}(t),\ 1 \le j \le \delta_i - \gamma_i,\ v_i(t), \\
&\quad i \in (1, \dots, \rho^* \mid \xi_i = 0)\}, \{x^C_{ij}(t),\ 1 \le j \le \delta_i - \gamma_i + 1, \\
&\quad i \in (1, \dots, \rho^* \mid \xi_i > 0)\}, v^2(t)) \\
u^2(t) &= v^2(t)
\end{aligned}
$$

and the immersion map $Q(x, x^C)$ may be chosen as follows

$$
Q(x, x^C) = \begin{bmatrix} \psi^0_{01}(x) \\ \psi^*_{11}(x, x^C) \\ \vdots \\ \psi^*_{k1}(x, x^C) \\ x^C \end{bmatrix}.
$$

We want to stress that the dynamic state feedback that solves the problem of linearization the input-dependent part of the output, provides also the solution to the problem of immersion into a linear system, provided the assumption (B) is satisfied (see the sufficiency part of the proof).

Comparing the conditions of Theorem 6.15 and those of Theorem 6.23, we may conclude that the difference between solvability conditions for the problem of immersion into a linear system and that of linearization the input-dependent part of the output, is related to finite dimensionality (over the reals) of the space ψ of certain functions, obtained in the application of the inversion algorithm. Note that these functions ψ^0_k correspond to those output components which cannot be affected by inputs. So, the class of systems for which the two problems are equivalent consist of systems for which the condition (B) holds automatically. One such subclass is a subclass of right invertible systems, i.e systems with $\rho^* = p$. Really, the condition (B) of Theorem 6.23 holds for right invertible systems automatically since in case of right invertible systems $\psi^0_{\alpha+\lambda} \equiv 0$ for $\lambda \geq 0$. This observation generalizes the result of (Isidori and Ruberti 1985) which says that for continuous-time right-invertible systems, the problem of linearization the input-dependent part of the output and that of immersion into a linear system, via static state feedback, coincide.

Corollary 6.24 *For systems that are locally right invertible about a given trajectory the problem of immersion a system by regular dynamic state feedback into a linear system, is always solvable.*

Proof. For systems that are locally right invertible about the given trajectory the problem of linearizing the input-dependent part of the output is always solvable by Corollary 6.16. The fact that the condition (B) always holds for right invertible systems, concludes the proof. ∎

Theorem 6.25 *Consider the system S described by equations (5.1) around the regular trajectory $\{\bar{x}(t), \bar{u}(t), \bar{y}(t);\ 0 \le t \le t_F\}$ with respect to some specific application of the inversion algorithm. The system S is locally around the regular trajectory immersed by the Singh compensator into a linear system if and only if in case of this specific application of the inversion algorithm*

(A) for all integers $0 \le k \le t_F$, $\hat{y}_\alpha(k) := [y_{\rho^*+1}(k), \ldots, y_p(k)]^T$, obtained at the kth step of the inversion algorithm, i.e.

$$\hat{y}_\alpha(k) = \psi_{k2}(x(0), \{y_i(j),\ 1 \le i \le \rho_k,\ \gamma_i \le j \le k\}),$$

reduces to the following form

$$\hat{y}_\alpha(k) = \psi_k^0(x(0), \{y_i(j),\ 1 \le i \le \rho_k,\ \gamma_i \le j \le \delta_i - 1\}) + \sum_{i=1}^{\rho_k} \sum_{j=\delta_i}^{k} a_{ij}^k y_i(j),$$

where the elements of $(p - \rho^*)$-dimensional vectors a_{ij}^k are constants and

(B) dim ψ *is finite, where ψ is the vector space (over R) generated by functions $\psi_k^0, k \ge 0$,*

$$\psi = \text{span}_R\{\psi_0^0, \psi_1^0, \ldots, \psi_k^0, \ldots\}.$$

The proof of Theorem 6.25 is similar to though technically more complicated than the proof of Theorem 6.23 and will be omitted.

Now, let us compare our necessary and sufficient conditions with the result by Lee and Marcus (1988). For this purpose let us reformulate their result in terms of our (more general) inversion algorithm.

Theorem 6.26 *Consider the system S, described by equations (5.1) around the regular trajectory with respect to some specific application of the inversion algorithm. The system S is locally around the regular trajectory immersed by regular static state feedback into a linear system if and only if in case of this specific application of the inversion algorithm*

(A') for all integers $0 \le k \le t_F$ the $\hat{y}_k(k)$, obtained at the kth step of in the inversion algorithm, i.e.

$$\hat{y}_k(k) = \psi_k(x(0),\ \{y_i(j),\ 1 \le i \le \rho_k,\ \gamma_i \le j \le k\})$$

reduces to the following special form

$$\hat{y}_k(k) = \sum_{i=1}^{k} \sum_{j=i}^{k} K_{ij}^k \tilde{y}_i(j) + b_k(x(0)),$$

where the elements of $(p - \rho_k) \times (\rho_i - \rho_{i-1})$-dimensional matrices K_{ij}^k are constants and

(B') $\dim \psi$ is finite, where

$$\psi = \text{span}_R\{b_{\alpha+j}(x),\ j \geq 0\}.$$

Comparing $(A), (B)$ with $(A'), (B')$ we may conclude that solvability of the dynamic immersion problem does not necessarily imply that the static immersion problem is solvable. Of course, if the latter problem is solvable, then is also the dynamic immersion problem.

Our procedure for constructing the compensator unfortunately leads us to a dynamic compensator even if a static solution exists. That is, the compensator (6.23) does not necessarily reduce to static state feedback, if a static immersion problem is solvable. However, we can check condition (A') and (B') without extra computations during the first part of our procedure, i.e. during applying the inversion algorithm to system (5.1). Then, if these conditions are satisfied, we can instead of the compensator (6.23) construct the static state feedback in correspondence with (Lee and Marcus 1988). Note, that if the condition (A') is satisfied, then equation (6.19) reduces to

$$\begin{aligned}
\tilde{y}_1(t+1) &= \tilde{a}_1^*(x(t), u(t)) \\
&\dots \\
\tilde{y}_\alpha(t+\alpha) &= \sum_{i=1}^{\alpha-1} \sum_{j=i+1}^{\alpha} K_{ij}^\alpha \tilde{y}_i(t+j) + \tilde{a}_\alpha^*(x(t), u(t)).
\end{aligned} \tag{6.52}$$

Lee and Marcus (1988) obtained a static state feedback compensator $u(t) = \varphi(x(t), v(t))$ as a solution of the following system of equations

$$\begin{aligned}
v^1(t) &= \begin{bmatrix} \tilde{a}_1^*(x(t), u(t)) \\ \dots \\ \tilde{a}_\alpha^*(x(t), u(t)) \end{bmatrix} \\
v^2(t) &= u^2(t).
\end{aligned} \tag{6.53}$$

Now we shall present various examples that illustrate our theory.

Example 6.27 (Continuation of Example 6.17)

Applying the first order dynamic compensator (6.39) as suggested in Section 6.5 we obtain the closed-loop system

$$\begin{aligned}
x_1(t+1) &= x^C(t) \\
x_2(t+1) &= x_3(t)[x^C(t) - x_2(t)] \\
x_3(t+1) &= \frac{v_2(t)}{v_1(t) - x_3(t)[x^C(t) - x_2(t)]} \\
x_4(t+1) &= x_4(t) + x^C(t) - x_2(t) \\
z(t+1) &= v_1(t) \\
y_1(t) &= x_1(t),\ y_2(t) = x_2(t),\ y_3(t) = x_4(t)
\end{aligned} \tag{6.54}$$

whose input-dependent part of the output is linear in the new control v.

Under the (extended) state-space transformation

$$x_1^* = x_1\,,\; x_2^* = x_2\,,\; x_3^* = x_4\,,\; x_4^* = x_3(x^C - x_2)\,,\; x_5^* = x_4 + x^C - x_2\,,$$

the system (6.54) is immersed into the following linear system

$$\begin{aligned} x_1^*(t+1) &= x_5^*(t) + x_2^*(t) - x_3^*(t) \\ x_2^*(t+1) &= x_4^*(t) \\ x_3^*(t+1) &= x_5^*(t) \\ x_4^*(t+1) &= v_2(t) \\ x_5^*(t+1) &= x_5^*(t) - x_4^*(t) + v_1(t) \\ y_1(t) &= x_1^*(t)\,,\; y_2(t) = x_2^*(t)\,,\; y_3(t) = x_3^*(t)\,. \end{aligned}$$

Note that for this system the condition (A') is not satisfied. So, there exist no static state feedback that solves the problem of immersion the system into a linear system.

Example 6.28 Consider the system

$$\begin{aligned} x_1(t+1) &= x_2(t) + u_1(t) \\ x_2(t+1) &= x_3(t)u_1(t) \\ x_3(t+1) &= x_3(t)u_2(t) \\ x_4(t+1) &= x_5(t)x_4(t) + u_1(t) \\ x_5(t+1) &= x_5(t) \\ y_1(t) &= x_1(t)\,,\; y_2(t) = x_2(t)\,,\; y_3(t) = x_4(t)\,. \end{aligned}$$

Applying the inversion algorithm to this system, we obtain

$$\begin{aligned} y_1(t+1) &= x_2(t) + u_1(t) \\ y_2(t+2) &= x_3(t)u_2(t)[y_1(t+2) - x_3(t)u_1(t)] \end{aligned}$$

and

$$\begin{aligned} y_3(t+2) &= x_5^2(t)x_4(t) - x_2(t) - x_3(t)[y_1(t+1) - x_2(t)] + y_1(t+1) + y_1(t+2) \\ y_3(t+k) &= x_5^k(t)x_4(t) - x_2(t) - x_3(t)[y_1(t+1) - x_2(t)] + \\ &\quad + \sum_{j=1}^{k} y_1(t+j) - \sum_{j=2}^{k-1} y_2(t+j)\,,\; k \geq 3\,. \end{aligned}$$

So, $\alpha = 2$, $\gamma_1 = 1$, $\varepsilon_1 = 2$, $\gamma_2 = \varepsilon_2 = 2$.

Since the condition (A) of Theorem 6.23 (as well of Theorem 6.25) is satisfied, the input-dependent part of the output of the system is linearizable by dynamic compensator of the form (6.23) with arbitrary initial state $x^C(0) = x_0^C$ such that $x_0^C \neq x_{20} - v_1(0)/x_{30}$

$$x^C(t+1) = v_1(t)$$
$$u_1(t) = x^C(t) - x_2(t)$$
$$u_2(t) = \frac{v_2(t)}{x_3(t)\{v_1(t) - x_3(t)[x^C(t) - x_2(t)]\}}\,.$$

Applying this compensator, it is easy to see that the input-dependent parts of the output components are linear in new control v and independent of initial state (x_0, x_0^C):

$$\begin{aligned}
&y_1(t+1) = x^C(t)\,,\ y_1(t+2) = v_1(t)\,,\\
&y_2(t+1) = x_3(t)[x^C(t) - x_2(t)]\,,\ y_2(t+2) = v_2(t)\,,\\
&y_3(t+1) = x_5(t)x_4(t) + x^C(t) - x_2(t)\,,\\
&y_3(t+2) = x_5^2(t)x_4(t) - x_2(t) - x_3(t)[x^C(t) - x_2(t)] + x^C(t) + v_1(t)\,,\\
&y_3(t+k) = x_5^k(t)x_4(t) - x_2(t) - x_3(t)[x^C(t) - x_2(t)] + x^C(t) +\\
&\qquad + \sum_{j=2}^{k} v_1(t+j-2) - \sum_{j=2}^{k-1} v_2(t+j-2)\,,\ k \ge 3\,.
\end{aligned}$$

The condition (B) of Theorem 6.23 for this system is not satisfied since $\dim \psi$, where

$$\psi = \mathrm{span}_{\mathcal{R}}\{x^C, x_1, x_2, x_4, x_3(x^C - x_2), x_5x_4 + x^C - x_2,\ x_5x_4 + x^C - x_2 - x_3(x^C - x_2), \ldots, x_5^k x_4 + x^C - x_2 - x_3(x^C - x_2), \ldots\}$$

is not finite.

Example 6.29 (Continuation of Example 6.19)
Adding time-shifts into the compensator C changes the vector space ψ, generated by $\psi_k^0(x, x^C)$, $0 \le k \le t_F$. Note that in case of C, $\psi_2^0(x, x^C) = x_5x_3(x_1^C - x_2)$, whereas in case of $C1$, $\psi_2^0(x, x^C) = x_5[x_2^C - x_3(x_1^C - x_2)]$.

In order to solve the problem of immersion a system considered in this example into a linear system by feedback $C1$ on the time interval $0 \le t \le 2$,

$$\psi_2^0(x, x^C) = x_5[x_2^C - x_3(x_1^C - x_2)]$$

must depend linearly on the components of $\psi_0^0(x) = [x_1, x_2, x_4]$, $\psi_1^0(x, x^C) = [x_3(x_1^C - x_2), x_5(x_1^C - x_2)]$ and x^C, which is of course impossible since the term $x_5x_2^C$.

Consider now the time interval $0 \le t \le 3$ and compensator $C2$.

In order to solve the problem of immersion an original system into a linear system by feedback $C2$ on the time interval $0 \le t \le 3$, $\psi_3^0(x, x^C) = x_5[x_3^C - x_4^C]$ must depend linearly on the components of $\psi_0^0(x) = [x_1, x_2, x_4]$, $\psi_1^0(x, x^C) = [x_3(x_1^C - x_2), x_5(x_1^C - x_2)]$, $\psi_2^0(x, x^C) = x_5x_2^C - x_5x_3(x_1^C - x_2)$ and x^C, which is of course impossible since $\psi_3^0(x, x^C)$ depends on x_3^C and x_4^C.

Notes and References

The local model matching problem formulation for discrete-time nonlinear systems (Definition 6.1) reminds the corresponding problem formulation for continuous-time systems [MPC91]. In the MMP the requirement that $y^{S\circ C} - y^M$ does not depend on u^M amounts, in the linear case, to the equality of the transfer functions of the compensated system and of the model. In the nonlinear case, the so-called higher order Volterra kernels of $S \circ C$ and M have to coincide. Note that in the given formulation of the MMP, the zero-th order Volterra kernels of $S \circ C$ and M, which represent the responses to the input $u^M(t) = u^{M0}$, $0 \leq t \leq t_F$, are not required to be the same, thus following a practice in use for the linear MMP.

The proof on Lemma 6.2 may be found in [Kot92a]; it is a discrete-time analogue of the one given in [MPC91].

The solution of the MMP is due to Kotta [Kot93]; the solvability conditions (see Theorem 6.3) are in full accordance with the corresponding conditions for linear systems [Mal82], and are the same as those for continuous-time nonlinear systems [MPC91]. However, it has been claimed in [MPC91] that the above conditions are only sufficient and not necessary; an example due to Huijberts [Hui92] has been presented to confirm the argument. It has been shown later [Kot94b] that under slightly stronger regularity assumptions on the equilibrium point than those of [MPC91], the conditions by Moog, Perdon and Conte are also necessary and that the example due to Huijberts treats precisely the situation when these slightly stronger regularity assumptions are not satisfied.

The input-output linearization problem via static state feedback has been solved in [LM87]. The here presented solution is different in order to be able to compare the solvability conditions via static and dynamic compensators.

The Singh compensator has been introduced by Huijberts [Hui91] for the continuous-time nonlinear systems.

The input-output decoupling problem has been solved by Nijmeijer [Nij87, Nij90]. However, the solution presented here is different since Nijmeijer did not use the inversion algorithm to construct the compensator but its dual version, the so-called extension algorithm [Nij87]. In [Nij90] it is shown that under generic conditions the problems of dynamic input-output decoupling and right invertibility for discrete-time nonlinear systems are solvable around an equilibrium point if and only if the same problems are solvable for the linearization of the nonlinear system. For differential geometric conditions for solvability of dynamic I/O decoupling problem, see [Gri86], [CG92].

The results on input-output linearization problem and immersion a nonlinear system into a linear system, via dynamic state feedback may be found in [KN94a] and [KN94b], respectively. For the results on immersion problem via static state feedback, see [LM88].

[BI86] Di Benedetto M.D. and A.Isidori. The matching of nonlinear models via dynamic state feedback. *SIAM J. Control and Optimization*, 1986, v. 24, 1063–1075.

[CG92] Chung S.T. and J.W.Grizzle. Internally exponentially stable non-linear discrete-time non-interacting control via static feedback. *Int. J. Contr.*, 1992, v. 55, 1071–1092.

[Fli87] Fliess M. Esquisses pour une théorie des systemes non linéaires en temps discret. Rend. Sem. Mat. Univ. Politec. Torino. Fasc. Spec. Control Theory, 1987, 55–67.

[Gri86] Grizzle J.W. Local input-output decoupling of discrete-time nonlinear systems. *Int. J. Control*, 1986, v. 43, 1517–1530.

[Hui91] Huijberts H. Dynamic feedback in nonlinear synthesis problems. Ph.D. Thesis, University of Twente, Enschede, 1991.

[Hui92] Huijberts H.J.C. A nonregular solution of the nonlinear dynamic disturbance decoupling problem vith an application to a complete solution of the nonlinear model matching problem. *SIAM J. Control and Optimization*, 1992, v. 30, 350–366.

[Kot92a] Kotta Ü. Model matching of nonlinear discrete-time systems in the presence of unmeasurable disturbances. *Prepr. 2nd IFAC Symp. on Nonlinear Control Systems Design*. Bordeaux, 1992, 563–568.

[Kot92b] Kotta Ü. Model matching of nonlinear discrete-time systems in the presence of measurable disturbances. *Prepr. 11th Int. Conf. on Systems Science*, Wroclaw, Poland, 1992.

[Kot93] Kotta Ü. Structural approach to discrete-time nonlinear model matching. *Prepr. of 12th IFAC World Congress*, Sydney, 1993, v. 9, 261–264.

[Kot94a] Kotta Ü. Model matching of nonlinear discrete-time systems in the presence of disturbances. *Proc. Estonian Acad. Sci. Phys. Math.*, 1994, v. 43, 7–14.

[Kot94b] Kotta Ü. Comments on a structural approach to nonlinear model matching problem. *SIAM J. Control and Optimization*, 1994, v. 32, 1555–1558.

[KN94a] Kotta Ü. and H.Nijmeijer. On dynamic input-output linearization of discrete-time nonlinear systems. *Int. J. Control*, 1994, v. 60, 1319–1337.

[KN94b] Kotta Ü. and H.Nijmeijer. Immersion a discrete-time nonlinear system by regular dynamic state feedback into a linear system. Submitted for publication.

[LM87] Lee H.G. and S.I.Marcus. On input-output linearization of discrete-time nonlinear systems. *Systems and Control Letters*, 1987, v. 8, 249–259.

[LM88] Lee H.G. and S.I.Marcus. Immersion and immersion by nonsingular feedback of a discrete-time nonlinear system into a linear ststem. *IEEE Trans. on Autom. Control*, 1988, v. 33, 479–483.

[Mal82] Malabre M. Structure a l'infini des triplets invariants. Application a la poursuite parfaite de modele. *Lect. Notes in Control and Inf. Sci.*, 1982, v. 44, 43–53.

[MNCI89] Monaco S., D.Normand-Cyrot and T.Isola. Nonlinear decoupling in discrete systems. *Proc. IFAC Symp. on Nonlinear Control Systems Design*, Capri, Italy, 1989, 48–55.

[MPC91] Moog C.H., A.M.Perdon and G.Conte. Model matching and factorization for nonlinear systems: a structural approach. *SIAM J. Control and Optimization*, 1991, v. 29, 769–785.

[Nij87] Nijmeijer H. Local (dynamic) input-output decoupling of discrete-time nonlinear systems. *IMA Journal of Mathematical Control & Information*, 1987, v. 4, 237–250.

[Nij90] Nijmeijer H. Remarks on the control of discrete-time nonlinear systems. In: *Perspectives in Control Theory.* (Eds.) B.Jakubczyk, K.Malanowski, W.Respondek. Birkhäuser, Boston, 1990, 261–276.

[NS90] Nijmeijer H. and A. van der Schaft. *Nonlinear Dynamical Control Systems.* Berlin, Springer-Verlag, 1990.

7. Systems with Input Disturbances. The General Case

This chapter studies the model matching problem for a discrete time nonlinear system in the presence of disturbances and the dynamic disturbance decoupling problem. The problems are treated locally around an equilibrium point of the system and both the cases of measurable and unmeasurable disturbances are considered. The complete solutions to the considered problems, that is necessary and sufficient conditions as well the procedures for constructing the compensator, are given under certain regularity assumptions of the equilibrium point. Instrumental in the solution of these problems are the two versions of the inversion (structure) algorithm for the discrete time nonlinear system with disturbances which produce two finite sequences of uniquely defined integers, the so-called invertibility indices, either with respect to control, or with respect to both inputs (See Chapter 5). Necessary and sufficient conditions for local solvability of the considered problems are derived in terms of the invertibility indices of three systems: the original system, the so-called extended system, formed by the original system and the model, and the auxiliary system which has obtained from the original system by delaying its input. If these conditions are satisfied, the inversion algorithm provides the systematic procedures for constructing the compensators that solve the considered problems.

The organization of the chapter is as follows. In the following section the mathematical formulation of the model matching problems in the presence of disturbances, either measurable or unmeasurable, will be given. In sections 7.2 and 7.3 we give our main results on MMP for the cases of measurable and unmeasurable disturbances respectively. In section 7.4 the formulation of the dynamic disturbance decoupling problem will be given, again both for the cases of measurable and unmeasurable disturbances. The sections 7.5 and 7.6 provide the solutions to these problems for the cases of measurable and unmeasurable disturbances, respectively.

7.1 The Formulation of the Model Matching Problem in the Presence of Disturbances

Consider a discrete-time nonlinear plant S, described by equations (4.1), i.e. the equations of the form

$$\begin{aligned} x(t+1) &= \ (x(t), u(t), w(t)), x(0) = x_0\,, \\ y(t) &= h(x(t))\,, \end{aligned} \tag{7.1}$$

where $x(t), u(t), w(t), y(t), f$ and h are defined as before.

Furthermore, let a discrete-time nonlinear model M be given:

$$\begin{aligned} x^M(t+1) &= f^M(x^M(t), u^M(t)), x^M(0) = x_0^M\,, \\ y^M(t) &= h^M(x^M(t))\,, \end{aligned} \tag{7.2}$$

where again $x^M(t), u^M(t), y^M(t), f^M$ and h^M are defined as before (See Section 3.1).

In the case of measurable disturbances, the compensator C_w used to control the system S is a discrete-time nonlinear system described by

$$\begin{aligned} x^C(t+1) &= f^C(x^C(t), x(t), u^M(t), w(t)), x^C(0) = x_0^C\,, \\ u(t) &= h^C(x^C(t), x(t), u^M(t), w(t)) \end{aligned} \tag{7.3}$$

with the state $x^C(t) \in X^C$, an open subset of R^{n_C}, and real analytic f^C and h^C. When the disturbances are not available for measurement, the equations of compensator cannot depend on w and then the compensator C is described by

$$\begin{aligned} x^C(t+1) &= f^C(x^C(t), x(t), u^M(t)),\ x^C(0) = x_0^C\,, \\ u(t) &= h^C(x^C(t), x(t), u^M(t)) \end{aligned} \tag{7.4}$$

with the state $x^C(t) \in X^C$, an open subset of R^{n_C} and real analytic f^C and h^C.

The composition of (7.1) and (7.3)

$$\begin{aligned} x(t+1) &= f(x(t), h^C(x^C(t), x(t), u^M(t), w(t)), w(t)) \\ x^C(t+1) &= f^C(x^C(t), x(t), u^M(t), w(t)) \\ y^{S \circ C}(t) &= h(x(t))\,, \end{aligned} \tag{7.5}$$

initialized at (x_0, x_0^C) is denoted by $S \circ C_w$; the composition of (7.1) and (7.4)

$$\begin{aligned} x(t+1) &= f(x(t), h^C(x^C(t), x(t), u^M(t)), w(t)) \\ x^C(t+1) &= f^C(x^C(t), x(t), u^M(t)) \\ y^{P \circ C}(t) &= h(x(t))\,, \end{aligned} \tag{7.6}$$

initialized at (x_0, x_0^C) is denoted by $S \circ C$.

We assume to work in a neighbourhood of an equilibrium point of the system (7.1), that is around $(x^0, u^0, w^0) \in X \times U \times W$ such that $f(x^0, u^0, w^0) = x^0$. The equilibrium point (x^{M0}, u^{M0}) of the model M is said to correspond to the equilibrium point (x^0, u^0, w^0) of the system S if the following equalities hold:

$$\begin{aligned} f^M(x^{M0}, u^{M0}) &= x^{M0} \\ h^M(x^{M0}) &= y^{M0} = y^0 = h(x^0)\,. \end{aligned}$$

Again, we say that the equilibrium point $(x^{C0}, x^0, u^{M0}, w^0, u^0)$ of the compensator (7.3) corresponds to the equilibrium points (x^0, u^0, w^0) and (x^{M0}, u^{M0}) of the system S and the model M, if the following equalities hold:

$$f^C(x^{C0}, x^0, u^{M0}, w^0) = x^{C0}\,, \quad h^C(x^{C0}, x^0, u^{M0}, w^0) = u^0\,.$$

The correspondence of the equilibrium point $(x^{C0}, x^0, u^{M0}, u^0)$ of the compensator (7.4) to the equilibrium points of S and M is defined analogously:

$$f^C(x^{C0}, x^0, u^{M0}) = x^{C0}\,, \quad h^C(x^{C0}, x^0, u^{M0}) = u^0\,.$$

Definition 7.1 Local model matching problem in the presence of measurable disturbances. *Given the system S around an equilibrium point (x^0, u^0, w^0), the model M around an equilibrium point (x^{M0}, u^{M0}) corresponding to (x^0, u^0, w^0) and a point (x_0, x_0^M), find if possible, a compensator C_w of the form (7.3) together with an initial state x_0^C and neighbourhoods $V_1 = X^{C0} \times X^0 \times U^{M0} \times W^0$ of $(x^{C0}, x^0, u^{M0}, w^0)$ in $X^C \times X \times U^M \times W$ and V_2 of u^0 in U, being domain and range of C_w, as well a neighbourhood X^{M0} of x^{M0} and a map $\xi : X^{M0} \to X^{C0}$ with the property that*

$$y^{S \circ C_w}(t, x_0, \xi(x_0^M), w(0), \ldots, w(t-1), u^M(0), \ldots, u^M(t-1)) - \\ - y^M(t, x_0^M, u^M(0), \ldots, u^M(t-1))\,, \; 0 \le t \le t_F\,,$$

does not depend on u^M and w for all $(x_0, x_0^M) \in X^0 \times X^{M0}$, for all w and u^M in the domain of C_w.

Definition 7.2 Local model matching problem in the presence of unmeasurable disturbances. *Given the system S around an equilibrium point (x^0, u^0, w^0), the model M around an equilibrium point (x^{M0}, u^{M0}) corresponding to (x^0, u^0, w^0) and a point (x_0, x_0^M), find if possible, a compensator C defined by equations of the form (7.4) together with an initial state x_0^C and neighbourhoods $V_1 = X^{C0} \times X^0 \times U^{M0}$ of (x^{C0}, x^0, u^{M0}) in $X^C \times X \times U^M$ and V_2 of u^0 in U, being domain and range of C, as well a neighbourhood X^{M0} of x^{M0} and a map $\xi : X^{M0} \to X^{C0}$ with the property that*

$$y^{S \circ C}(t, x_0, \xi(x_0^M), w(0), \ldots, w(t-1), u^M(0), \ldots, u^M(t-1)) - \\ - y^M(t, x_0^M, u^M(0), \ldots, u^M(t-1))\,, \; 0 \le t \le t_F\,,$$

does not depend on u^M and w for all $(x_0, x_0^M) \in X^0 \times X^{M0}$, for all u^M in the domain of C, and for all w around w^0.

7.2 The Solution of the Model Matching Problem in the Presence of Measurable Disturbances

We now give necessary and sufficient conditions for the local solvability of the MMP for S and M in the presence of measurable disturbances. For this purpose, like in Section 6.1 we introduce the extended system SM, associated with the system S and the model M:

$$\begin{aligned} x(t+1) &= f(x(t), u(t), w(t)) \\ x^M(t+1) &= f^M(x^M(t), u^M(t)) \\ y^{SM}(t) &= h^{SM}(x(t), x^M(t)) = h(x(t)) - h^M(x^M(t)) \,. \end{aligned} \tag{7.7}$$

The extended system (7.7) can be viewed as the system of the form (7.1) with control $u(t)$, with measurable disturbances $w(t)$ and $u^M(t)$, and with equilibrium point $(x^0, x^{M0}, u^0, w^0, u^{M0})$.

Theorem 7.3 *Consider the system S described by equations (7.1) around an equilibrium point (x^0, u^0, w^0) and the model M described by equations (7.2) around an equilibrium point (x^{M0}, u^{M0}), corresponding to (x^0, u^0, w^0). Assume that the equilibrium point of the system SM is strongly regular with respect to both versions of the inversion algorithm. The local model matching problem in the presence of measurable disturbances for S and M is solvable, if and only if*

$$\rho_{uwu^M,i}(SM) = \rho_{u,i}(S), \; i \geq 1 \,. \tag{7.8}$$

Proof. Sufficiency. Assume that (7.8) holds. Then by Lemma 6.2 we have

$$\rho_{u,i}(SM) = \rho_{uwu^M,i}(SM) \,. \tag{7.9}$$

This implies that both versions of the inversion algorithm applied to system SM coincide and at the αth step we obtain

$$\begin{aligned} \tilde{y}_1^{SM}(t+1) &= \tilde{a}_1^{SM}(x(t), x^M(t), u(t), u^M(t), w(t)) \\ \tilde{y}_2^{SM}(t+2) &= \tilde{a}_2^{SM}(x(t), x^M(t), u(t), u^M(t), w(t), \tilde{y}_1^{SM}(t+2)) \\ &\dots \\ \tilde{y}_\alpha^{SM}(t+\alpha) &= a_\alpha^{SM}(x(t), x^M(t), u(t), u^M(t), w(t), \\ &\qquad \{\tilde{y}_i^{SM}(t+j), \, 1 \leq i \leq \alpha - 1, \, i+1 \leq j \leq \alpha\}) \end{aligned} \tag{7.10}$$

and

$$\hat{y}_\alpha^{SM} = \psi_\alpha^{SM}(x(t), x^M(t), \{\tilde{y}_i^{SM}(t+j), \, 1 \leq i \leq \alpha, \, i \leq j \leq \alpha\}) \,. \tag{7.11}$$

According to the inversion algorithm, the Jacobian matrix of the right hand side of (7.10) with respect to u around the point $(x^0, x^{M0}, u^0, u^{M0}, w^0)$ has full row rank ρ_u^*. Moreover, $\tilde{a}_i^{SM}(x^0, x^{M0}, u^0, u^{M0}, w^0), \{0, \dots, 0\}) = 0$, $i = 1, \dots, \alpha$. So we may solve equation (7.10) for $u(t)$ around the point $(x^0, x^{M0}, u^0, u^{M0}, w^0), \{0, \dots, 0\})$ by applying the Implicit Function Theorem. In order to get the simplest compensator we solve equation (7.10) under the assumption $\tilde{y}_i^{SM}(t+j) = 0$, $1 \leq i \leq \alpha$, $i \leq j \leq \alpha$:

$$u(t) = \varphi(x(t), x^M(t), u^M(t), w(t)) \tag{7.12}$$

which is such that for $i = 1, \dots, \alpha$

$$\tilde{a}_i^{SM}(x(t), x^M(t), \varphi(x(t), x^M(t), u^M(t), w(t)), u^M(t), w(t), \{0, \dots, 0\}) = 0 \,. \tag{7.13}$$

Notice that $\varphi : V_1 \to V_2$ is analytic for some (possible small) neighbourhoods V_1 and V_2 of $(x^0, x^{M0}, u^{M0}, w^0)$ in $X \times X^M \times U^M \times W^0$ and u^0 in U^0. This implies that (7.13) will be hold as long as $(x(t), x^M(t), u^M(t), w(t)) \in V_1$ and $u(t) \in V_2$. Of course, the equality (7.13) is lost if we leave the neighbourhoods V_1, or V_2.

Construct the compensator C_w as

$$\begin{aligned} x^C(t+1) &= f^M(x^C(t), u^M(t)),\ x^C(0) = x_0^M\,, \\ u(t) &= \varphi(x(t), x^M(t), u^M(t), w(t))\,. \end{aligned} \tag{7.14}$$

We proceed to show that (7.14) and $\xi = \mathrm{Id}$ (identity map) serve as the solution of the MMP in the presence of measurable disturbances. Our proof starts with the observation that by (7.9), $\hat{y}_\alpha^{SM}(t+k)$ in (7.11) remains independent from $u^M(t)$ and $w(t)$ for all $k > \alpha$, since otherwise $\rho_{uwu^M,k}(SM)$ would be strictly greater than $\rho_{u,k}(SM)$. Therefore, $\hat{y}_k^{S\circ C}(t+k) - \hat{y}_k^M(t+k)$ is independent from u^M and w for every $k \geq 1$. It is immediate from (7.13) that

$$\tilde{y}_j^{S\circ C}(t+j) = \tilde{y}_j^M(t+j)\,,\ j = 1, \ldots, \alpha\,.$$

Thus, $y^{S\circ C}(t) - y^M(t)$ is independent from u^M and w for $0 \leq t \leq t_F$ which completes the sufficiency part of the proof.

Necessity. Let us assume that there exists a compensator C_w of the form (7.3) for S and M that locally around a strongly regular (with respect to both versions of the inversion algorithm) equilibrium point of SM solves the MMP. Apply the first step of the inversion algorithm to SM with respect to the control u only, considering disturbances w and u^M as parameters

$$\begin{aligned} \tilde{y}_1^{SM}(t+1) &= \tilde{a}_1^{SM}(x(t), x^M(t), u(t), u^M(t), w(t)) \\ \hat{y}_1^{SM}(t+1) &= \psi_1^{SM}(x(t), x^M(t), u^M(t), w(t), \tilde{y}_1^{SM}(t+1))\,, \end{aligned} \tag{7.15}$$

where

$$\mathrm{rank}\,\frac{\partial}{\partial u}\tilde{a}_1^{SM}(\cdot) = \rho_{u,1}(SM)\,.$$

Substituting the output of C_w in (7.15), we can assert that the equations do not depend on u^M and w any more since C_w solves the MMP for S and M. In particular, this means that either

$$\frac{\partial \psi_1^{SM}(\cdot)}{\partial(w, u^M)} \equiv 0 \tag{7.16}$$

everywhere around the point $(x^0, x^{M0}, u^{M0}, w^0, \tilde{y}_1^{SM,0})$, or, if not, the compensator C_w will guarantee equality (7.16). The latter is impossible around the strongly regular equilibrium point where by definition $\partial\psi_1^{SM}/\partial(w, u^M)$ is either equal to zero everywhere or different from zero everywhere. This means that if $\partial\psi_1^{SM}/\partial(w, u^M) \not\equiv 0$, we can never make it equal to zero by the suitable choice of compensator. This implies that $\partial\psi_1^{SM}/\partial(w, u^M) \equiv 0$ which gives us

$$\rho_{uwu^M,1}(SM) = \mathrm{rank}\begin{bmatrix} \partial\tilde{a}_1^{SM}/\partial u & \partial\tilde{a}_1^{SM}/\partial(w, u^M) \\ 0 & \partial\psi_1^{SM}/\partial(w, u^M) \end{bmatrix} = \mathrm{rank}\,\frac{\partial}{\partial u}\tilde{a}_1^{SM} = \rho_{u,1}(SM)\,.$$

Applying this argument repeatedly, we finally get $\rho_{uwu^M,1}(SM) = \rho_{u,i}(SM)$, $i \geq 1$. The conclusion of the necessity part of the theorem follows using Lemma 6.2.

∎

7.3 The Solution of the Model Matching Problem in the Presence of Unmeasurable Disturbances

We now give necessary and sufficient conditions for the local solvability of the MMP for S and M in the presence of unmeasurable disturbances. For this purpose we introduce the auxiliary system S_a formed from the original system S by adding one-step backward shift (delay) operator into the control loop. So, the auxiliary system S_a with the states $(x(t), u(t))$ and the inputs $u_a(t)$ is defined as follows

$$\begin{aligned} x(t+1) &= f(x(t), u(t), w(t)) \\ u(t+1) &= u_a(t) \\ y(t) &= h(x(t)). \end{aligned} \tag{7.17}$$

The equilibrium point of S_a is (x^0, u^0, u_a^0, w^0) with $u_a^0 = u^0$.
First we prove several lemmas.

Lemma 7.4 *Apply the inversion algorithm to S with respect to u around an equilibrium point (x^0, u^0, w^0). Suppose that the point (x^0, u^0, w^0) is strongly regular equilibrium point of S with respect to both versions of the inversion algorithm. Then at every step of the algorithm $\partial\psi_k(\cdot)/\partial w = 0$ if and only if $\rho_{uw,k}(S) = \rho_{u,k}(S)$ for all k.*

Proof. Let us consider in detail the first step of the inversion algorithm. Compute $y(t+1) = h(f(x(t), u(t), w(t)))$ and define

$$\begin{aligned} \rho_{u,1} &= \text{rank} \frac{\partial}{\partial u} h(f(x,u,w))\Big|_{x=x^0, u=u^0, w=w^0}, \\ \rho_{uw,1} &= \text{rank} \frac{\partial}{\partial(u,w)} h(f(x,u,w))\Big|_{x=x^0, u=u^0, w=w^0}. \end{aligned}$$

Now permute the components of the output so that the first $\rho_{u,1}$ rows of $h(f(x,u,w))/\partial u$ are linearly independent and write accordingly

$$\begin{aligned} \tilde{y}_1(t+1) &= \tilde{a}_1(x,u,w) \\ \hat{y}_1(t+1) &= \hat{a}_1(x,u,w), \end{aligned} \tag{7.18}$$

where $\partial\tilde{a}_1(x,u,w)/\partial u$ has full row rank $\rho_{u,1}$. Then from (7.18) we have $u = \xi(x, w, \tilde{y}_1(t+1))$ for some ξ and hence $\hat{y}_1(t+1) = \hat{a}_1(x, \xi(x, w, \tilde{y}_1(t+1)), w) = \psi_1(x, w, \tilde{y}_1(t+1))$. Moreover, from the identity $\tilde{y}_1 = \tilde{a}_1(x, \xi(x, w, \tilde{y}_1), w)$ we obtain

$$\frac{\partial\tilde{a}_1}{\partial u}\frac{\partial\xi}{\partial w} + \frac{\partial\tilde{a}_1}{\partial w} = 0 \quad \text{or} \quad \frac{\partial\xi}{\partial w} = -\left(\frac{\partial\tilde{a}_1}{\partial u}\right)^+ \frac{\partial\tilde{a}_1}{\partial w},$$

where $(\partial\tilde{a}_1/\partial u)^+$ is a right inverse of $\partial\tilde{a}_1/\partial u$, that is $(\partial\tilde{a}_1/\partial u)(\partial\tilde{a}_1/\partial u)^+ = I$. Now, assume that

$$\frac{\partial\psi_1(t+1)}{\partial w} = \frac{\partial\hat{a}_1}{\partial w} + \frac{\partial\hat{a}_1}{\partial u}\frac{\partial\xi}{\partial w} = \frac{\partial\hat{a}_1}{\partial w} - \frac{\partial\hat{a}_1}{\partial u}\left(\frac{\partial\tilde{a}_1}{\partial u}\right)^+\frac{\partial\tilde{a}_1}{\partial w} = 0\,.$$

Taking into account that $\partial\hat{a}_1/\partial u = \alpha(x,u,w)[\partial\tilde{a}_1/\partial u]$, we can easily check that this yields

$$A^* = \begin{bmatrix} \dfrac{\partial\tilde{a}_1}{\partial u} & \dfrac{\partial\tilde{a}_1}{\partial w} \\ \dfrac{\partial\hat{a}_1}{\partial u} & \dfrac{\partial\hat{a}_1}{\partial w} \end{bmatrix} = \begin{bmatrix} I_{\rho_{u,1}} \\ \dfrac{\partial\hat{a}_1}{\partial u}\left(\dfrac{\partial\tilde{a}_1}{\partial u}\right)^+ \end{bmatrix} \begin{bmatrix} \dfrac{\partial\tilde{a}_1}{\partial u}\,\dfrac{\partial\tilde{a}_1}{\partial w} \end{bmatrix},$$

$$\rho_{uw,1} = \operatorname{rank} A^* = \rho_{u,1}\,.$$

Conversely, assume that $\rho_{u,1} = \rho_{uw,1}$. This implies that there is a matrix $\alpha(x,u,w)$ such that

$$\frac{\partial\hat{a}_1}{\partial u} = \alpha\frac{\partial\tilde{a}_1}{\partial u}\,,\quad \frac{\partial\hat{a}_1}{\partial w} = \alpha\frac{\partial\tilde{a}_1}{\partial w}\,.$$

Hence

$$\frac{\partial\hat{a}_1}{\partial w} - \frac{\partial\hat{a}_1}{\partial u}\left(\frac{\partial\tilde{a}_1}{\partial u}\right)^+\frac{\partial\tilde{a}_1}{\partial w} = \alpha\frac{\partial\tilde{a}_1}{\partial w} - \alpha\frac{\partial\tilde{a}_1}{\partial u}\left(\frac{\partial\tilde{a}_1}{\partial u}\right)^+\frac{\partial\tilde{a}_1}{\partial w} = 0$$

and thus

$$\frac{\partial\psi_1(\cdot)}{\partial w} = \frac{\partial\hat{a}_1}{\partial w} - \frac{\partial\hat{a}_1}{\partial u}\left(\frac{\partial\tilde{a}_1}{\partial u}\right)^+\frac{\partial\tilde{a}_1}{\partial w} = 0\,.$$

Applying the above arguments repeatedly, we prove the lemma. ∎

Lemma 7.5 *Apply the inversion algorithm with respect to control u around a strongly regular equilibrium point (x^0, u^0, w^0) to S. Let the point (x^0, u^0, w^0) be such that (x^0, u^0, u_a^0, w^0) with $u_a^0 = u^0$ is a strongly regular equilibrium point of S_a with respect to both versions of the inversion algorithm. Then at every step of the algorithm $\partial a_k^S(\cdot)/\partial w = 0$ if and only if*

$$\rho_{u_a w,k}(S_a) = \rho_{u_a,k}(S_a) \text{ for all } k \geq 1\,.$$

Proof. Let us consider the first step of the inversion algorithm applied to S and S_a. Compute

$$y(t+1) = h(f(x(t),u(t),w(t))) = a_1^S(x,u,w) = a_1^{S_a}(x,u,w)$$

and define

$$\begin{aligned} \rho_{u,1}(S) &= \operatorname{rank}\frac{\partial}{\partial u}a_1^S(\cdot)\,,\\ \rho_{u_a,1}(S_a) &= \operatorname{rank}\frac{\partial}{\partial u_a}a_1^{S_a}(\cdot)\,,\\ \rho_{u_a w,1}(S_a) &= \operatorname{rank}\frac{\partial}{\partial(u_a,w)}a_1^{S_a}(\cdot)\,. \end{aligned}$$

Assume that $\partial a_1^S(\cdot)/\partial w = 0$. As $a_1^S(\cdot) = a_1^{S_a}(\cdot)$, this yields

$$\rho_{u_a w,1}(S_a) = \rho_{u_a,1}(S_a).$$

Conversely, assume that $\rho_{u_a w,1}(S_a) = \rho_{u_a,1}(S_a)$. Take into account that by construction of the system S_a, $\rho_{u_a,1}(S_a) = 0$; this implies that $\partial a_1^{S_a}(\cdot)/\partial w = \partial a_1^S(\cdot)/\partial w = 0$. So, $\partial a_1^S(\cdot)/\partial w = 0$ iff $\rho_{u_a w,1}(S_a) = \rho_{u_a,1}(S_a)$.

Permute the components of the outputs of both systems S and S_a so that the first $\rho_{u,1}$ rows of $\partial a_1^S(\cdot)/\partial u$ are linearly independent and write accordingly (taking into account that either $\partial a_1^S(\cdot)/\partial w = 0$, or equivalently $\rho_{u_a w,1}(S_a) = \rho_{u_a,1}(S_a)$):

$$\begin{aligned} \tilde{y}_1^S(t+1) &= \tilde{a}_1^S(x,u) \\ \hat{y}_1^S(t+1) &= \hat{a}_1^S(x,u), \end{aligned} \tag{7.19}$$

where $\partial \tilde{a}_1^S(\cdot)/\partial u$ has full row rank $\rho_{u,1}(S)$. Then from (7.19) we have $u(t) = \xi(x(t), \tilde{y}_1^S(t+1))$ and hence $\hat{y}_1^S(t+1) = \hat{a}_1^S(x(t), \xi(x(t), \tilde{y}_1^S(t+1)))$. Moreover, from the identity $\tilde{y}_1^S(t+1) \equiv \tilde{a}_1^S(x(t), \xi(x(t), \tilde{y}_1^S(t+1)))$ we obtain

$$\frac{\partial \tilde{a}_1^S}{\partial x} + \frac{\partial \tilde{a}_1^S}{\partial u}\frac{\partial \xi}{\partial x} = 0$$

or

$$\frac{\partial u}{\partial x} = -\left(\frac{\partial \tilde{a}_1^S}{\partial u}\right)^+ \frac{\partial \tilde{a}_1^S}{\partial x}, \tag{7.20}$$

where $(\partial \tilde{a}_1^S/\partial u)^+$ is a right inverse of $\partial \tilde{a}_1^S/\partial u$, that is $(\partial \tilde{a}_1^S/\partial u)(\partial \tilde{a}_1^S/\partial u)^+ = I$.

Now, consider in detail the second step of the inversion algorithm for S and S_a. Compute

$$\hat{y}_1^S(t+2) = \hat{a}_1^S(f(x,u,w), \xi(f(x,u,w), \tilde{y}_1^S(t+2))) = a_2^S(x,u,w,\tilde{y}_1^S(t+2)) \tag{7.21}$$

and

$$\hat{y}_1^{S_a}(t+2) = \begin{bmatrix} \tilde{a}_1^S(f(x,u,w),u_a) \\ \hat{a}_1^S(f(x,u,w),u_a) \end{bmatrix} \triangleq \begin{bmatrix} a_{21}^{S_a}(x,u,w,u_a) \\ a_{22}^{S_a}(x,u,w,u_a) \end{bmatrix} = a_2^{S_a}(x,u,w,u_a). \tag{7.22}$$

From (7.22) we obtain

$$\rho_{u_a,2}(S_a) = \operatorname{rank}(\partial a_2^{S_a}(\cdot)/\partial u_a) = \operatorname{rank}\frac{\partial}{\partial u} a_1^S(\cdot) = \rho_{u,1}(S). \tag{7.23}$$

Assume that $\partial a_2^S(\cdot)/\partial w = 0$. As

$$\frac{\partial a_2^S}{\partial w} \overset{\text{by(7.21)}}{=} \frac{\partial \hat{a}_1^S}{\partial x}\frac{\partial f}{\partial w} + \frac{\partial \hat{a}_1^S}{\partial u}\frac{\partial \xi}{\partial x}\frac{\partial f}{\partial w} \overset{\text{by(7.20)}}{=} \frac{\partial \hat{a}_1^S}{\partial x}\frac{\partial f}{\partial w} - \frac{\partial \hat{a}_1^S}{\partial u}\left(\frac{\partial \tilde{a}_1^S}{\partial u}\right)^+ \frac{\partial \tilde{a}_1^S}{\partial x}\frac{\partial f}{\partial w} =$$

$$\overset{\text{by(7.22),(7.23)}}{=} \frac{\partial a_{22}^{S_a}}{\partial w} - \frac{\partial a_{22}^{S_a}}{\partial u_a}\left(\frac{\partial a_{21}^{S_a}}{\partial u_a}\right)^+ \frac{\partial a_{21}^{S_a}}{\partial w}, \tag{7.24}$$

this implies

$$A^* = \frac{\partial}{\partial(u_a, w)} a_2^{S_a}(\cdot) = \begin{pmatrix} I_{\rho_{u_a,2}}(S_a) \\ \frac{\partial a_{22}^{S_a}}{\partial u_a} \left(\frac{\partial a_{21}^{S_a}}{\partial u_a}\right)^+ \end{pmatrix} \left(\frac{\partial a_{21}^{S_a}}{\partial u_a} \; \frac{\partial a_{21}^{S_a}}{\partial w}\right).$$

Thus

$$\rho_{u_a w,2}(S_a) = \text{rank}\, A^* = \rho_{u_a,2}(S_a).$$

Conversely, assume that $\rho_{u_a w,2}(S_a) = \rho_{u_a,2}(S_a)$. This implies that there is a matrix $\alpha(x, u, w, u_a)$ such that

$$\frac{\partial a_{22}^{S_a}}{\partial u_a} = \alpha \frac{\partial a_{21}^{S_a}}{\partial u_a}, \quad \frac{\partial a_{22}^{S_a}}{\partial w} = \alpha \frac{\partial a_{21}^{S_a}}{\partial w}.$$

Hence

$$\frac{\partial a_{22}^{S_a}}{\partial w} - \frac{\partial a_{22}^{S_a}}{\partial u_a} \left(\frac{\partial a_{21}^{S_a}}{\partial u_a}\right)^+ \frac{\partial a_{21}^{S_a}}{\partial w} = 0,$$

and thus by (7.24), $\partial a_2^S(\cdot)/\partial w = 0$. Applying the above arguments repeatedly, we prove the lemma. ∎

Now we are ready to prove our main result.

Theorem 7.6 *Consider the system S described by equations (7.1) around an equilibrium point (x^0, u^0, w^0) and the model M described by equations (7.2) around an equilibrium point (x^{M0}, u^{M0}), corresponding to (x^0, u^0, w^0). Assume that the equilibrium points of the systems S, S_a and SM are strongly regular with respect to all versions of the inversion algorithm considered in the formulation of the theorem. The local model matching problem in the presence of unmeasurable disturbances for S and M is solvable, if and only if*

$$\rho_{u_a w,i}(S_a) = \rho_{u_a,i}(S_a), \; i \geq 1 \tag{7.25}$$

and

$$\rho_{u u^M,i}(SM) = \rho_{u,i}(S), \; i \geq 1. \tag{7.26}$$

Proof. Sufficiency. By (7.25) and by Lemma 6.2 we have

$$\rho_{u_a w,i}(S_a M) = \rho_{u_a,i}(S_a M), \; i \geq 1,$$

and therefore by Lemma 7.5

$$\partial a_k^{SM}/\partial w = 0, \; k \geq 1. \tag{7.27}$$

Furthermore, assume that (7.26) holds. Then by Lemma 6.2

$$\rho_{u u^M,i}(SM) = \rho_{u,i}(SM), \; i \geq 1$$

and therefore by Lemma 7.4

$$\partial\psi_k^{SM}(\cdot)/\partial u^M = 0\,, \; k \geq 1\,. \tag{7.28}$$

Now, apply the inversion algorithm with respect to control u to SM. By (7.27) and (7.28) we obtain at the last step of the algorithm

$$\begin{aligned} \tilde{y}_1^{SM}(t+1) &= \tilde{a}_1^{SM}(x(t), x^M(t), u(t), u^M(t)) \\ \tilde{y}_2^{SM}(t+2) &= \tilde{a}_2^{SM}(x(t), x^M(t), u(t), u^M(t), \tilde{y}_1^{SM}(t+2)) \\ &\dots \\ \tilde{y}_\alpha^{SM}(t+\alpha) &= \tilde{a}_\alpha^{SM}(x(t), x^M(t), u(t), u^M(t), \\ &\qquad \{\tilde{y}_i^{SM}(t+j),\; 1 \leq i \leq \alpha-1,\; i+1 \leq j \leq \alpha\}) \end{aligned} \tag{7.29}$$

and

$$\hat{y}_\alpha^{SM}(t+\alpha) = \psi_\alpha^{SM}(x(t), x^M(t), \{\tilde{y}_i^{SM}(t+j),\; 1 \leq i \leq \alpha,\; i \leq j \leq \alpha\})\,. \tag{7.30}$$

The rest of the proof may be handled in much the same way as the proof of Theorem 7.3, the only difference is that unlike equations (7.10), equations (7.29) do not depend on $w(t)$. Therefore, the solution of (7.29), i.e. $u(t) = \varphi(x(t), x^M(t), u^M(t))$, does not depend on w either.

Necessity. Let us assume that there exists a compensator C of the form (7.5) for S and M that locally solves the MMP. Assume at first that the condition (7.25) does not hold for $i = 1$, that is

$$\rho_{u_a w,1}(S_a) \neq \rho_{u_a,1}(S_a)\,.$$

By Lemma 7.5 this means that $\partial a_1^S(\cdot)/\partial w \neq 0$. Then at the first step of the inversion algorithm $y^{SM}(t+1)$ explicitly depends on w

$$y^{SM}(t+1) = h(x(t+1)) - h^M(x^M(t+1)) = a_1^S(x(t), u(t), w(t)) - a_1^M(x^M(t), u^M(t))\,. \tag{7.31}$$

Since (7.5) solves the MMP for S and M, this w-dependence should disappear, if we plug (7.5) into (7.31). Since (7.5) does not depend on w, this is not possible, except the case if (7.5) is such that it imposes the constraint

$$\frac{\partial a_1^S(\cdot)}{\partial w} = 0\,. \tag{7.32}$$

Of course, the latter is not possible around the regular equilibrium point of S_a. The reason is that around the regular equilibrium point $\partial a_1^S(\cdot)/\partial w$ is everywhere either equal to zero or different from zero. This means that if $\partial a_1^S(\cdot)/\partial w \neq 0$, we can never make if equal to zero by the suitable choice of the compensator. So we necessarily have that (7.25) holds for $i = 1$. Applying the same arguments repeatedly, we finally have that (7.25) holds for every $i \geq 1$.

Now, let assume that (7.26) does not hold for $i = 1$, that is

$$\rho_{uu^M,1}(SM) \neq \rho_{u,1}(S)\,.$$

By Lemma 6.2 this means that

$$\rho_{uu^M,1}(SM) \neq \rho_{u,1}(SM).$$

Then applying the first step of the inversion algorithm to SM with respect to control u only, by Lemma 7.4 we obtain

$$\frac{\partial \psi_1^{SM}(\cdot)}{\partial u^M} \neq 0$$

and that

$$\hat{y}_1^{SM}(t+1) = \hat{a}_1^{SM}(x(t), x^M(t), u^M(t), w(t), \tilde{y}_1^{SM}(t+1)) \tag{7.33}$$

explicitly depends on u^M. Since (7.5) solves the MMP for S and M, this u^M-dependence should disappear, if we plug (7.5) into (7.33). But as (7.33) does not depend on u explicitly, this is not possible except (7.5) is such that it imposes the constraint

$$\frac{\partial \psi_1^{SM}(t+1)}{\partial u^M} = 0.$$

Of course, the latter is not possible around the regular equilibrium point of SM. So we necessarily have that (7.26) holds for $i = 1$. Applying the same arguments repeatedly, we finally have that (7.26) holds for every $i \geq 1$. ∎

7.4 The Formulation of the Dynamic Disturbance Decoupling Problem

In this section we shall formulate the disturbance decoupling problem with the dynamic state feedback. The problem deals with the situation in which by means of the dynamic state feedback compensator we want to achieve decoupling between the input disturbances entering the system and the outputs leaving the system. Clearly, the dynamic disturbance decoupling problem (DDDP) is essentially a nonlinear problem: for linear systems the DDDP is solvable if and only if the disturbance decoupling problem is solvable by static state feedback.

Consider a discrete time nonlinear system S, described by equations (7.1), i.e. by equations

$$\begin{aligned} x(t+1) &= f(x(t), u(t), w(t)),\ x(0) = x_0\,, \\ y(t) &= h(x(t))\,. \end{aligned}$$

We shall consider separately the cases of unmeasurable and measurable disturbances. In the case of unmeasurable disturbances the compensator C (dynamic state feedback) used to control the system is a discrete-time nonlinear system described by the equations of the form

$$\begin{aligned} x^C(t+1) &= f^C(x^C(t), x(t), v(t)), x^C(0) = x_0^C\,, \\ u(t) &= h^C(x^C(t), x(t), v(t))\,, \end{aligned} \tag{7.34}$$

with the state $x^C(t) \in X^C$, an open subset of R^{n_C} with a new m-dimensional control $v(t) \in V$, an open subset of R^m and real analytic f^C and h^C.

The *regularity* of (7.34) means that the dynamical system

$$\begin{aligned} x(t+1) &= f(x(t), h^C(x^C(t), x(t), v(t)), w(t)) \\ x^C(t+1) &= f^C(x^C(t), x(t), v(t)) \\ u(t) &= h^C(x^C(t), x(t), v(t)) \end{aligned} \tag{7.35}$$

with inputs $v(t)$ and outputs $u(t)$ defines a one-to-one (x, x^C, w)-dependent correspondence between the input variable v and output variable u.

The closed-loop system (7.1), (7.34), that is the system

$$\begin{aligned} x(t+1) &= f(x(t), h^C(x^C(t), x(t), v(t)), w(t)), \\ x^C(t+1) &= f^C(x^C(t), x(t), v(t)), \\ y(t) &= h(x(t)). \end{aligned} \tag{7.36}$$

initialized at (x_0, x_0^C) is denoted by $S \circ C$.

Definition 7.7 Local (regular) dynamic disturbance decoupling problem in the case of unmeasurable disturbances (DDDPud). *Given the system (7.1) around an equilibrium point (x^0, u^0, w^0), find, if possible, a (regular) compensator C defined by equations of the form (7.34) together with an initial state x_0^C and neighbourhoods $V_1 = X^{C0} \times X^0 \times V^0$ of (x^{C0}, x^0, v^0) in $X^C \times X \times V$ and V_2 of u^0 in U, being the domain and the range of C, as well a neighbourhood W^0 of w^0 so that the outputs of the closed-loop system*

$$y^{S \circ C}(t; x_0, x_0^C, w(0), \ldots, w(t-1), v(0), \ldots, v(t-1)), \ 0 \le t \le t_F$$

do not depend on disturbances $w(t)$ for every $x_0 \in X^0$, all $v(t) \in V^0$ and all $w(t) \in W^0$.

In the case of the measurable disturbances the compensator C_w (dynamic state feedback) used to control the system is a discrete-time nonlinear system described by the equations of the form

$$\begin{aligned} x^C(t+1) &= f^C(x^C(t), x(t), v(t), w(t)), \ x^C(0) = x_0^C, \\ u(t) &= h^C(x^C(t), x(t), v(t), w(t)) \end{aligned} \tag{7.37}$$

with the state $x^C(t) \in X^C \subset R^{n_C}$, with a new m-dimensional control $v(t) \in V \subset R^m$ and real analytic f^C and h^C. The r e g u l a r i t y of (7.37) means that the dynamical system

$$\begin{aligned} x(t+1) &= f(x(t), h^C(x^C(t), x(t), v(t), w(t)), w(t)), \\ x^C(t+1) &= f^C(x^C(t), x(t), v(t), w(t)), \\ u(t) &= h^C(x^C(t), x(t), v(t), w(t)) \end{aligned} \tag{7.38}$$

with inputs $v(t)$ and outputs $u(t)$ defines a one-to-one (x, x^C, w)-dependent correspondence between the input variable v and output variable u.

The closed-loop system (7.1), (7.37), that is the system

$$\begin{aligned} x(t+1) &= f(x(t), h^C(x^C(t), x(t), v(t), w(t)), w(t)), \\ x^C(t+1) &= f^C(x^C(t), x(t), v(t), w(t)), \\ y(t) &= h(x(t)) \end{aligned} \tag{7.39}$$

initialized at (x_0, x_0^C) is denoted by $S \circ C_w$.

Definition 7.8 Local (regular) dynamic disturbance decoupling problem in the case of the measurable disturbances (DDDPmd). *Given the system (7.1) around an equilibrium point (x^0, u^0, w^0), find, if possible, a (regular) compensator C_w defined by equations (7.37) together with an initial state x_0^C and neighbourhoods $V_1 = X^{C0} \times X^0 \times V^0 \times W^0$ of (x^{C0}, x^0, v^0, w^0) in $X^C \times X \times V \times W$ and V_2 of u^0 in U, being the domain and the range of C_w, as well a neighbourhood W^0 of w^0 so that the outputs of the closed-loop system*

$$y^{S \circ C_w}(t; x_0, x_0^C, w(0), \dots, w(t-1), v(0), \dots, v(t-1)), \; 0 \le t \le t_F$$

do not depend on disturbances $w(t)$ for every $x_0 \in X^0$, for all $v(t) \in V^0$, and for all $w(t) \in W^0$.

7.5 The Dynamic Disturbance Decoupling Problem Solution: the Case of Unmeasurable Disturbances

Performing the Inversion Algorithm at system (7.1) gives in each step $0 \le k \le n-1$ a function ψ_k representing the functionally dependent part of the output which does not depend on control. Considering the effects of applying the compensator to the output components, it will come as no surprise that the ψ_k's will play a crusial role in solving the DDDP.

At first, necessary and sufficient conditions are given in terms of functions ψ_k, $0 \le k \le n-1$, appearing in the inversion algorithm.

Theorem 7.9 *Apply the inversion algorithm to S, described by equations (7.1), with respect to the control u around a strongly regular equilibrium point (x^0, u^0, w^0). Then the DDDPud for system S is locally solvable around (x^0, u^0, w^0) via regular dynamic state feedback if and only if for $0 \le k \le n-1$*

$$\frac{\partial}{\partial w} \psi_k^S(f(x,u,w), \{\tilde{y}_i(t+j+1), \; 1 \le i \le k, \; i \le j \le k\}) = 0 . \tag{7.40}$$

Moreover, if (7.40) holds, the DDDPud can be solved around (x^0, u^0, w^0) by means of the Singh compensator (6.23) with arbitrary initial state.

Proof. Sufficiency. If (7.40) holds, then applying the inversion algorithm to (7.1) with $w = w^0$ gives the same result as applying it to (7.1) where we consider w as a parameter. Therefore, the compensator (6.23) applied to (7.1) locally around (x^0, u^0, w^0) yields

$$y_i(\gamma_i + j - 1) = x_{ij}^C(0)\,, \qquad j = 1, \dots, \varepsilon_i - \gamma_i$$
$$y_i(t + \varepsilon_i) = v_i(t)\,, \qquad 0 \le t \le t_F\,,\ i = 1, \dots, \rho^*$$

which are, of course, independent from $w(t)$. Furthermore, if (7.40) holds for $0 \le k \le n-1$, it will also hold for every $k > n-1$. Therefore,

$$y_i(j) \text{ for } (1 \le i \le \rho^*,\ 0 \le j \le \gamma_i - 1) \text{ and } (\rho^* + 1 \le i \le p,\ j \ge 0)$$

being the components of $\hat{y}_k(t+k)$, $k \ge 0$, are independent from $w(t)$ by assumption. Therefore, (6.23) solves the DDDPud.

Necessity. Let us assume that there exists a regular dynamic state feedback defined by (7.37) for (7.1) that locally solves the DDDPud. Furthermore, assume that (7.40) does not hold for $k = 0$, that is

$$\frac{\partial h(f(x(t), u(t), w(t)))}{\partial w} \neq 0\,.$$

Then at the first step of the inversion algorithm we have that $y(t+1)$ explicitly depends on w:

$$y(t+1) = h(f(x(t), u(t), w(t)))\,. \tag{7.41}$$

Since (7.37) solves the DDDPud for (7.1), this w-dependence should disappear if we substitute (7.37) in (7.41). However, this is not possible since (7.37) does not depend on w. Thus (7.37) must be such that it imposes the constraint

$$\frac{\partial h(f(x(t), u(t), w(t)))}{\partial w} \triangleq \xi_1(x(t), u(t), w(t)) = 0\,,$$

i.e. that there exists at least one component of $u(t)$ which does not depend on the new control v.

The latter would imply the nonregularity of the compensator. So we necessarily have that (7.40) holds for $k = 0$.

Next, assume that (7.40) does not hold for $k = 1$. Then we obtain at the second step of the inversion algorithm (where we consider w as a parameter):

$$\frac{\partial \psi_1(f(x(t), u(t), w(t)), \tilde{y}_1(t+2))}{\partial w} \neq 0\,.$$

Using the same argument as above, we can see that this w-dependence will not disappear, unless (7.37) is constructed in such a way that the constraint

$$\frac{\partial \psi_1(f(x(t), u(t), w(t)), \tilde{y}_1(t+2))}{\partial w} \triangleq \xi_2(x(t), u(t), w(t), \tilde{y}_1(t+2)) = 0$$

is imposed on the system which would contradict the regularity of (7.37). Therefore, (7.40) has to hold for $k = 1$. Following the same way, we can show that (7.40) holds for $k = 0, 1, \dots, n-1$. ■

One may ask whether Theorem 7.9 is still true if we replace 'regular dynamic state feedback' by 'dynamic state feedback'. The next example demonstrates that this is not the case.

Example 7.10 Consider the system

$$\begin{aligned} x_1(t+1) &= x_2(t)u(t) \\ x_2(t+1) &= x_3(t)+w(t) \\ x_3(t+1) &= x_1(t)u(t) \\ y_1(t) = x_1(t),\ & y_2(t) = x_2(t)\,. \end{aligned}$$

It is easy to see that we can solve the DDDP for this system by $u = 0$ while confition (7.40) is not satisfied.

Now we are going to translate necessary and sufficient conditions for solvability of the DDDPud via regular compensator, formulated by Theorem 7.9 in terms of the inversion algorithm, into system-intrinsic and algorithm-independent conditions stated in terms of the invertibility indices $\rho_{u_a w,k}$ and $\rho_{u_a,k}$ of the auxiliary system S_a, defined by equations (7.17). In order to be able to define the invertibility indices of S_a, we have to work under slightly stronger regularity assumption than that assumed in Theorem 7.9. (See Section 5.7). Namely, we must assume strong regularity with respect to *both versions* of the inversion algorithm and not just with respect to the second version (i.e. inversion with regard to control).

Theorem 7.11 *Consider the system S described by equations (7.1) around a strongly regular equilibrium point (x^0, u^0, w^0) with respect to the inversion algorithm with regard to the control. Let the point (x^0, u^0, w^0) be such that (x^0, u^0, u_a^0, w^0) with $u_a^0 = u^0$ is a strongly regular equilibrium point of S_a with respect to both versions of the inversion algorithm. Then the regular DDDPud for S is locally solvable around (x^0, u^0, w^0) if and only if for all $1 \le k \le n+m$*

$$\rho_{u_a w,k}(S_a) = \rho_{u_a,k}(S_a)\,. \tag{7.42}$$

Proof. The proof of Theorem 7.11 follows easily from Theorem 7.9 and Lemma 7.5. ∎

The conclusion is not affected if we replace 'regular DDDPud' by 'DDDPud' in Theorem 7.11.

Theorem 7.12 *Consider the system S described by equations (7.1) around a strongly regular equilibrium point (x^0, u^0, w^0) with respect to the inversion algorithm with regard to control. Let the point (x^0, u^0, w^0) be such that (x^0, u^0, u_a^0, w^0) with $u_a^0 = u^0$ is a strongly regular equilibrium point of S_a with respect to both versions of the inversion algorithm. Then the DDDPud for S is locally solvable if and only if (7.42) holds.*

Proof. Sufficiency. As in Theorem 7.9.

Necessity. Let us assume that there exists a dynamic feedback control in the form (7.34) for (7.1) that locally around a strongly regular equilibrium point (x^0, u^0, w^0)

solves the DDDPud. Apply the first step of the inversion algorithm to S with respect to control u. Then we obtain

$$y(t+1) = h(f(x(t), u(t), w(t)))\,. \tag{7.43}$$

If we plug (7.34) in (7.43), the equation does not depend on w any more since (7.34) solves the DDDPud for (7.1). This means that either

$$\frac{\partial h(f(x, u, w))}{\partial w} = 0 \tag{7.44}$$

everywhere around the point (x^0, u^0, w^0) or, if not, the compensator (7.34) will guarantee the equality (7.44). Note that around the strongly regular equilibrium point (x^0, u^0, u_a^0, w^0) of S_a, $\partial h(f(x, u, w))/\partial w$ is everywhere either equal to zero or different from zero. This means that if $\partial h(f(x, u, w))/\partial w \neq 0$ we can never make it equal to zero by the suitable choice of the compensator. This implies that (7.44) holds which by Lemma 7.5 means that also (7.42) holds for $k = 0$. Applying this argument repeatedly, we can show that (7.42) holds for k $k = 0, 1, \ldots, n-1$. ∎

Theorems 7.11 and 7.12 may be summarized by saying that nothing can be gained from nonregular compensators if we work around a strongly regular equilibrium point with respect to both versions of the inversion algorithm. Around a nonregular equilibrium point with respect to the first version of the inversion algorithm (i.e. inversion with respect to both inputs), the conditions (7.40) are not necessary for the solvability of the DDDPud. Sometimes, if the conditions (7.40) are not satisfied, a *nonregular* compensator can still be found that imposes the constraint (7.40) and solves the DDDPud.

7.6 The Dynamic Disturbance Decoupling Problem Solution: the Case of Measurable Disturbances

In this section we shall consider the case of measurable disturbances. We shall prove the following theorem.

Theorem 7.13 *Apply the inversion algorithm to (7.1) with respect to the control u around a strongly equilibrium point (x^0, u^0, w^0). Then the regular DDDPmd for system (7.1) is locally solvable around (x^0, u^0, w^0) if and only if for* $1 \leq k \leq n$

$$\frac{\partial}{\partial w}\psi_k(x, w, \{\tilde{y}_i(t+j),\ 1 \leq i \leq k,\ i \leq j \leq k\}) = 0\,. \tag{7.45}$$

Proof. Sufficiency. Notice at first that if (7.45) holds for $1 \leq k \leq n$, then it holds for every $k \geq 0$ since the inversion algorithm terminates in at most n steps. If (7.45) holds, then applying the second version of the inversion algorithm to (7.1) gives at the nth step

$$\tilde{Y}_n = A_n(x(t), u(t), w(t), \{\tilde{y}_i(t+j),\ 1 \le i \le n-1,\ i+1 \le j \le n\}) \quad (7.46)$$
$$\hat{y}_n(t+n) = \psi_n(x(t), \{\tilde{y}_i(t+j),\ 1 \le i \le n,\ i \le j \le n\}) \quad (7.47)$$

where $\tilde{Y}_n = [\tilde{y}_1^T(t+1), \tilde{y}_2^T(t+2), \ldots, \tilde{y}_n^T(t+n)]^T$ and the matrix $\partial A_n(\cdot)/\partial u$ has full row rank $\rho_{u,n}$ on a neighbourhood O_n of (x^0, u^0, w^0). Like in Section 6.3, equation (7.46) can be solved for $u^1(t)$ locally by applying the Implicit Function Theorem

$$u^1(t) = \varphi(x(t), \{y_i(t+j), 1 \le i \le \rho^*, \gamma_i \le j \le \varepsilon_i\}, u^2(t), w(t)) . \quad (7.48)$$

Now construct the compensator for (7.1) in the following way. The dynamic part of the compensator coincides with that of the Singh compensator (6.23) and the output equation of the compensator is defined as follows

$$u^1(t) = \varphi(x(t), \{x_{ij}^C(t+j),\ 1 \le j \le \varepsilon_i - \gamma_i,\ v_i(t),\ 1 \le i \le \rho^*\}, v^2(t), w(t)) ,$$
$$u^2(t) = v^2(t) . \quad (7.49)$$

It can be shown that the compensator described above is regular on a neighbourhood of (x^0, u^0, y^0, w^0). We omit the proof which is quite analogous to the case with unmeasurable disturbances.

Now, it is easy to see that the above compensator with arbitrary initial state, applied to (7.1) locally around (x^0, u^0, w^0) yields for $i = 1, \ldots, \rho_n$

$$y_i(\gamma_i + j - 1) = x_{ij}^C(0) , \quad j = 1, \ldots, \varepsilon_i - \gamma_i ,$$
$$y_i(t + \varepsilon_i) = v_i(t) , \quad 0 \le t \le t_F$$

which of course are independent from $w(t)$. Moreover, $y_i(j)$ for $1 \le i \le \rho^*$, $0 \le j \le \gamma_i - 1$ and for $\rho^* + 1 \le i \le p$, $j \ge 0$, being the components of $\hat{y}_k(k)$, do not depend on $w(t)$ by assumption. Hence, the compensator (6.23), (7.49) solves the regular DDDPmd locally.

Necessity. Let us assume, that there exists a regular dynamic feedback control defined by (7.37) for (7.1) that locally around the equilibrium point (x^0, u^0, w^0) solves the regular DDDPmd. Furthermore, assume that (7.45) does not hold for $k = 1$, that is

$$\frac{\partial}{\partial w}\psi_1(x, w, \tilde{y}_1(t+1)) \neq 0 .$$

Then at the first step of the inversion algorithm we have that $\hat{y}_1(t+1)$ explicitly depends on w:

$$\hat{y}_1(t+1) = \psi_1(x(t), w(t), \tilde{y}_1(t+1)) . \quad (7.50)$$

Since (7.37) solves the DDDPmd for (7.1) this w-dependence should disappear, if we use the compensator (7.37). Since (7.50) does not depend on control, the only possibility is that (7.37) must be such that it imposes the constraint

$$\frac{\partial}{\partial w}\psi_1(x, w, \tilde{y}_1(t+1)) = \xi_1(x, w, \tilde{a}_1(x, u)) = 0 .$$

But this would imply the nonregularity of the compensator. So, (7.45) must hold for $k = 1$. Following the same way, we prove the theorem. ∎

Note that the result of Theorem 7.13 provides us a constructive procedure for solving the regular DDDPmd. Namely, we proceed by applying the second version of the inversion algorithm to (7.1), checking at every step $1 \leq k \leq n$ if (7.45) holds. If (7.45) does not hold for some k, we conclude that the regular DDDPmd is not solvable. If (7.45) does hold for $1 \leq k \leq n$, then the regular DDDPmd can be solved by means of the compensator (6.23), (7.49) with arbitrary initial state.

Now, we are going to translate necessary and sufficient conditions for solvability of the regular DDDPmd, formulated by Theorem 7.13 in terms of the inversion algorithm into system-intrinsic and algorithm-independent conditions.

Theorem 7.14 *Consider the system S, described by equations (7.1) around a strongly regular equilibrium point (x^0, u^0, w^0), associated with both versions of the inversion algorithm. Then the regular DDDPmd for system S is locally solvable around (x^0, u^0, w^0) if and only if for $i \leq k \leq n$*

$$\rho_{uw,k}(S) = \rho_{u,1}(S)\,. \tag{7.51}$$

Proof. The proof of Theorem 7.14 follows easily from Theorem 7.13 and Lemma 7.5. ∎

Again, like in Section 7.5, if we work around a strongly regular equilibrium point with respect to both versions of the inversion algorithm, the conclusition is not affected if we replace 'regular DDDPmd' by 'DDDPmd' in Theorem 7.14.

Theorem 7.15 *Consider the system S described by equations (7.1) around a strongly regular equilibrium point (x^0, u^0, w^0), associated with both versions of the inversion algorithm. Then the DDDPmd for system S is locally solvable around (x^0, u^0, w^0) if and only if (7.51) holds for $i \leq k \leq n$.*

Proof. Sufficiency. As in Theorem 7.13.
Necessity. Similar to the proof of Theorem 7.12. ∎

7.7 An example

In this section we show explicitly for a particular nonlinear economic model that it is disturbance decouplable via dynamic state feedback. Consider the following model of a closed economy:

$$Y(t+1) = Y(t) + \alpha\{C(Y(t)) + I(Y(t), R(t), K(t)) + P^{-1}(t)G(t) - Y(t)\} \tag{7.52}$$
$$R(t+1) = R(t) + \beta\{L(Y(t), R(t)) - P^{-1}(t)M(t)\} \tag{7.53}$$
$$K(t+1) = K(t) + I(Y(t), R(t), K(t) \tag{7.54}$$
$$W^*(t+1) = \tilde{W}(t) \tag{7.55}$$
$$\tilde{W}(t+1) = W(t) \tag{7.56}$$

$$Y(t) = F(N(t), K(t)) \tag{7.57}$$
$$N(t) = H(W^*(t), P(t)) \,. \tag{7.58}$$

In this model the quantities have the following interpretation:

Y = real output;
C = real private consumption;
I = real private net investment;
R = nominal interest rate;
K = real capital stock;
P = price level;
G = nominal government spending;
L = real money demand;
M = nominal money stock;
N = labour demand;
W = nominal wage rate.

In (7.52) and (7.53) α and β are given positive constants. In this model, G and M are the control variables, W is a known disturbance variable, Y and P are the variables-to-be-controlled (the outputs).

We first have to transform (7.52)–(7.58) into the state space form (1.1); that is we have to rewrite (7.57)–(7.58). We will do this around a particular equilibrium point of (7.52)–(7.58), say $(\bar{Y}, \bar{R}, \bar{K}, \bar{W}, \bar{G}, \bar{M}, \bar{N})$. Consider the relation $N = H(W, P)$ which holds at the equilibrium point $(\bar{N}, \bar{W}, \bar{P})$. Provided that

$$\frac{\partial H}{\partial P}\Big|_{W,P} \neq 0 \,, \quad \frac{\partial F}{\partial N}\Big|_{N,K} \neq 0 \tag{7.59}$$

we may locally apply the Implicit Function Theorem yielding the equations

$$P = \tilde{H}(W^*, N) \,, \quad N = \tilde{F}(Y, K) \,,$$

and thus we obtain as the second output equation

$$P(t) = \tilde{H}(W^*(t), \tilde{F}(Y(t), K(t))) \,. \tag{7.60}$$

Altogether we have obtained – locally – a model of the form (7.1)

$$\begin{aligned} Y(t+1) &= f_1(Y(t), R(t), K(t), W^*(t), G(t)) \\ R(t+1) &= f_2(Y(t), R(t), K(t), W^*(t), M(t)) \\ K(t+1) &= f_3(Y(t), R(t), K(t)) \\ W^*(t+1) &= \tilde{W}(t) \\ \tilde{W}(t+1) &= W(t) \\ Q_1(t) &= Y(t) \\ Q_2(t) &= \tilde{H}(W^*(t), \tilde{F}(Y(t), K(t))) \end{aligned} \tag{7.61}$$

where Q_1 and Q_2 are the outputs (the variables-to-be-controlled) and the functions f_1, f_2 and f_3 directly follow from (7.52)–(7.54) and (7.60).

Applying the first step of the inversion algorithm to system (7.61), we obtain

$$Q_1(t+1) = f_1(Y(t), R(t), K(t), W^*(t), G(t)) \tag{7.62}$$

and

$$\begin{aligned} Q_2(t+1) &= \tilde{H}(W^*(t+1), \tilde{F}(Y(t+1), K(t+1)) = \\ &\triangleq \psi(\tilde{W}(t), Y(t), R(t), K(t), W^*(t), Q_1(t+1)) . \end{aligned}$$

At the second step of the algorithm we compute

$$\begin{aligned} Q_2(t+2) &= \psi(\tilde{W}(t+1), Y(t+1), R(t+1), K(t+1), W^*(t+1), Q_1(t+2)) = \\ &\triangleq \xi(W(t), Y(t), R(t), K(t), W^*(t), \tilde{W}(t), G(t), M(t), Q_1(t+2)) . \end{aligned} \tag{7.63}$$

By the Implicit Function Theorem the system of equations (7.62), (7.63) is locally around the equilibrium point solvable with respect to $G(t)$ and $M(t)$,

$$\begin{bmatrix} G(t) \\ M(t) \end{bmatrix} = \gamma(Y(t), R(t), K(t), W^*(t), \tilde{W}(t), W(t), Q_1(t+1), Q_1(t+2), Q_2(t+2))$$

if and only if the following conditions hold

$$\frac{\partial f_1}{\partial G}\Big|_{\bar{Y}, \bar{R}, \bar{K}, \bar{W}, \bar{G}} \neq 0 , \tag{7.64}$$

$$\frac{\partial \tilde{H}}{\partial \tilde{F}}\Big|_{\bar{W}, \bar{N}} \frac{\partial \tilde{F}}{\partial K}\Big|_{\bar{Y}, \bar{K}} \frac{\partial f_3}{\partial R}\Big|_{\bar{Y}, \bar{R}, \bar{K}} \frac{\partial f_2}{\partial M}\Big|_{\bar{Y}, \bar{R}, \bar{K}, \bar{W}, \bar{M}} \neq 0 . \tag{7.65}$$

By the structure of f_1, (7.64) holds. By the structure of f_2, $\partial f_2/\partial M \neq 0$ at the equilibrium point. Moreover, from (7.59) we obtain

$$\frac{\partial \tilde{H}}{\partial \tilde{F}}\Big|_{\bar{W}, \bar{N}} \neq 0$$

and so, in order to be able to solve the system of equations (7.62), (7.63) for G and M

$$\frac{\partial \tilde{F}}{\partial K}\Big|_{\bar{Y}, \bar{K}} \frac{\partial f_3}{\partial R}\Big|_{\bar{Y}, \bar{R}, \bar{K}} \neq 0 \tag{7.66}$$

has to be hold. Provided (7.66) holds, we can compute a disturbance decoupling control law

$$\begin{aligned} z(t+1) &= v_1(t) \\ \begin{bmatrix} G(t) \\ M(t) \end{bmatrix} &= \gamma(Y(t), R(t), K(t), W^*(t), \tilde{W}(t), W(t), z(t), v_1(t), v_2(t)) . \end{aligned}$$

Notes and References

The solution of the MMP in the presence of disturbances is presented in [Kot92b], [Kot92d], [Kot94a]. The problem has not been studied earlier for continuous-time nonlinear systems.

The regular DDDP for discrete-time systems has been solved locally around an equilibrium point in [KN91], [Kot92a], [Kot92c], [Kot94b]. The obtained results are actually discrete-time counterparts of results [HNW91], [HNW92] for continuous-time systems. For continuous-time nonlinear systems, the nonregular DDDP where the dynamic state feedback is not necessarily regular, is reported in [Hui92]. Up to now, no solution of the nonregular DDDP for discrete-time systems is available.

The example presented in Section 7.7, is taken from [KN91] and is actually a modification of the example, given in [NS90]. See [WK84] for an economic interpretation of (7.66).

In [FN94] the DDDP for discrete-time nonlinear systems is studied in a neighbourhood of a given trajectory. Furhtermore, in the above paper the connection between the solvability of this problem and the solvability of the corresponding problem for the time-varying linear system obtained by linearizing the original system along the given reference trajectory is investigated.

[FN94] Fliegner Th. and H.Nijmeijer. Dynamic disturbance decoupling for nonlinear discrete-time systems. *Proc. 33rd IEEE Conf. on Decision and Control*, Buena Vista, Florida, 1994, v. 2, 1790–1791.

[HNW91] Huijberts H.J.C., H.Nijmeijer, and L.L.M van der Wegen. Dynamic disturbance decoupling for nonlinear systems: the nonsquare and noninvertible case. In: *Controlled Dynamical Systems*, B.Bonnard, B.Bride, J.P.Gauthier and I.Kupka (Eds.), Boston, Birkhäuser, 1991, 243–252.

[HNW92] Huijberts H.J.C., H.Nijmeijer and L.L.M. van der Wegen. Dynamic disturbance decoupling for nonlinear systems. *SIAM J. Contr. and Optimization*, 1992, v. 30, 336–349.

[Hui92] Huijberts H.J.C. A nonregular solution of the nonlinear disturbance decoupling problem with an application to a complete solution of the nonlinear model matching problem. *SIAM J. Contr. and Optimization*, 1992, v. 30, 350–366.

[Kot92a] Kotta Ü. Dynamic disturbance decoupling for discrete-time nonlinear systems: the nonsquare and noninvertible case. *Proc. Estonian Acad. Sci. Phys. Math.*, 1992, v. 41, 14–22.

[Kot92b] Kotta Ü. Model matching of nonlinear discrete-time systems in the presence of unmeasurable disturbances. *Proc. IFAC Symp. on Nonlinear Control Systems Design*, Bordeaux, 1992, 563–568.

[Kot92c] Kotta Ü. Dynamic disturbance decoupling for discrete-time nonlinear systems: a solution in terms of system invariants. *Prepr. 2nd IFAC Workshop on System Structure and Control*, Prague, 1992, 200–203.

[Kot92d] Kotta Ü. Model matching of nonlinear discrete time systems in the presence of measurable disturbances. *Proc. 11th Int. Conf. on Systems Science*, Wroclaw 1992.

[Kot94a] Kotta Ü. Model matching of nonlinear discrete-time systems in the presence of disturbances. *Proc. Estonian Acad. Sci. Phys. Math.*, 1994, v. 43, 7–14.

[Kot94b] Kotta Ü. Dynamic disturbance decoupling for discrete-time nonlinear systems: a solution in terms of system invariants. *Proc. Estonian Acad. Sci. Phys. Math.*, 1994, v. 43, 147–159.

[KN91] Kotta Ü. and H.Nijmeijer. Dynamic disturbance decoupling for nonlinear discrete-time systems (In Russian). *Proc. of the Academy of USSR. Technical Cybernetics*, 1991, 52–57.

[NS90] Nijmeijer H. and A. van der Schaft. *Nonlinear Dynamical Control Systems*. Berlin, Springer-Verlag, 1990

[WK84] Wohltmann H.-W. and W.Krömer. Sufficient conditions for dynamic path controllability of economic systems. *Journal of Economic Dynamics and Control*, 1984, v. 7, 315–330.

Conclusions and Future Perspectives

The inversion method has numerous nice properties.

(i) It is straightforward and intuitively well understood.

(ii) The approach indicates how to set up the control objective so that a causal compensator that achieves this objective, exists.

(iii) After the inverse system has been found, compensators for different control problems can be easily designed with little extra computing.

(iv) The method can be used for a wide class of systems. Note that the application of the inversion method does not always require the complete invertibility of the system, sometimes the partial invertibility is enough.

However, a lot has remained to be done.

(i) The principal weakness of the method is that it assumes the complete knowledge of the system equations. To make the method more realistic, it should be combined with adaptive control methods. These questions are open for the future research. The preliminary results may be found in [BGC87] and [LNCM87] (see also [BS93], [SI89], [NA87], [BS91] for continuous-time systems).

(ii) At this point, one might feel that the problems of input-output linearization and decoupling, disturbance decoupling and model matching have been solved. However, one must remember that input-output maps only account for part of the closed-loop dynamics. Except the papers [Gri86], [CG92], [MNC87] the issues like what are the internal dynamics associated with the solutions of the above problems have been not treated for discrete-time nonlinear systems. If $\sum_{i=0}^{\alpha}(p-\rho_i) < n$, some state variables become unobserable after applying the feedback and the unobservable part (which is closely related to the reduced order right inverse system) has no reason to be stable, it stability depends nonlinearly on the choice of the reference trajectory. If the (reduced order) right inverse system is unstable, the control will be unbounded. One possibility is to replace the exact unstable inverse in (3.7) by a suitable stable approximate. These questions have been treated in [Ben73] for continuous-time linear systems. The stable inversion problem has recently been considered in [CP92] for a class of continuous-time nonlinear nonminimum phase systems with well defined relative degree whose zero dynamics has a hyperbolic equilibrium point.

A numerical procedure is also developed in [Che93] for constructing noncausal stable inverses based on iterative linearization and decomposition of the stable/unstable subspaces.

(iii) Finally, let us note that in [BJM93] the first attempts to consider invertibility and output dead-beat control for a clas of discrete-time nonlinear systems, described by input-output recursive representation, has been done.

[Ben73] Bengtsson G. A theory for control of linear multivariable systems. Lund Institute of Technology, Division of Automatic Control, 1973.

[BGC87] Bastin G., A.M.Guillaume, and G.Campion. A discrete-time self tuning computed torque controller for robotic manipulators. *Proc. of the 26th IEEE Conf. on Decision and Control*, Los Angeles, CA, 1987, 1026–1028.

[BJM93] Bastin G., F.Jarachi, I.M.Y.Mareels. Dead beat control of recursive nonlinear systems. *Proc. of the 32nd IEEE Conf. on Decision and Control*, 1993, 2965–2971.

[BS91] Di Benedetto M.D. and S.S.Sastry. Adaptive tracking for MIMO nonlinear systems. *Proc. ECC'91.*

[BS93] Di Benedetto M.D. and S.S.Sastry. Adaptive linearization and model reference control of a class of MIMO nonlinear systems, *J. of Mathematical Systems, Estimation and Control*, 1993, v. 3, 73–106.

[CG92] Chung S.T. and J.W.Grizzle. Internally exponentially stable non-linear discrete-time non-interacting control via static feedback. *Int. J. Contr.*, 1992, v. 55, 1071–1092.

[Che93] Chen D. An iterative solution to stable inversion of nonlinear nonminimum phase systems. *Proc. of American Control Conf.*, San Francisco, 1993.

[CP92] Chen D., Paden B. Stable inversion of nonlinear nonminimum phase systems. *Proc of Japan/USA Symp. on Flexible Automation*, 1992, 791–797.

[Gri86] Grizzle J.W. Local input-output decoupling of discrete-time nonlinear systems. *Int. J. Cont.*, 1986, v. 43, 1517–1530.

[LNCM87] Landau I.D., D.Normand-Cyrot and A.Montano. Adaptive control of a class of nonlinear discrete-time systems. Application to a heat exchanges. *Proc. of the 26th IEEE Conf. on Decision and Control*, Los Angeles, CA, 1987, 1990–1995.

[MNC87] Monaco S. and D.Normand-Cyrot. Minimum-phase nonlinear discrete-time systems and feedback stabilization. *Proc. 26th IEEE Conf. on Decision and Control*, Los Angeles, CA, 1987, 979-986.

[NA87] Nam K. and A.Arapostathis. A model reference adaptive control scheme for pure-feedback nonlinear systems. *IEEE Trans. Autom. Control*, 1988, v. 33, 803–811.

[SI89] Sastry S.S. and A.Isidori. Adaptive control of linearizable systems. *IEEE Trans. Autom. Control*, 1989, v. 34, 1123–1131.

Index

Lecture Notes in Control and Information Sciences

Edited by M. Thoma

1992–1995 Published Titles:

Vol. 167: Rao, Ming
Integrated System for Intelligent Control
133 pp. 1992 [3-540-54913-7]

Vol. 168: Dorato, Peter; Fortuna, Luigi; Muscato, Giovanni
Robust Control for Unstructured Perturbations: An Introduction
118 pp. 1992 [3-540-54920-X]

Vol. 169: Kuntzevich, Vsevolod M.; Lychak, Michael
Guaranteed Estimates, Adaptation and Robustness in Control Systems
209 pp. 1992 [3-540-54925-0]

Vol. 170: Skowronski, Janislaw M.; Flashner, Henryk; Guttalu, Ramesh S. (Eds)
Mechanics and Control. Proceedings of the 4th Workshop on Control Mechanics, January 21-23, 1991, University of Southern California, USA
302 pp. 1992 [3-540-54954-4]

Vol. 171: Stefanidis, P.; Paplinski, A.P.; Gibbard, M.J.
Numerical Operations with Polynomial Matrices: Application to Multi-Variable Dynamic Compensator Design
206 pp. 1992 [3-540-54992-7]

Vol. 172: Tolle, H.; Ersü, E.
Neurocontrol: Learning Control Systems Inspired by Neuronal Architectures and Human Problem Solving Strategies
220 pp. 1992 [3-540-55057-7]

Vol. 173: Krabs, W.
On Moment Theory and Controllability of Non-Dimensional Vibrating Systems and Heating Processes
174 pp. 1992 [3-540-55102-6]

Vol. 174: Beulens, A.J. (Ed.)
Optimization-Based Computer-Aided Modelling and Design. Proceedings of the First Working Conference of the New IFIP TC 7.6 Working Group, The Hague, The Netherlands, 1991
268 pp. 1992 [3-540-55135-2]

Vol. 175: Rogers, E.T.A.; Owens, D.H.
Stability Analysis for Linear Repetitive Processes
197 pp. 1992 [3-540-55264-2]

Vol. 176: Rozovskii, B.L.; Sowers, R.B. (Eds)
Stochastic Partial Differential Equations and their Applications. Proceedings of IFIP WG 7.1 International Conference, June 6-8, 1991, University of North Carolina at Charlotte, USA
251 pp. 1992 [3-540-55292-8]

Vol. 177: Karatzas, I.; Ocone, D. (Eds)
Applied Stochastic Analysis. Proceedings of a US-French Workshop, Rutgers University, New Brunswick, N.J., April 29-May 2, 1991
317 pp. 1992 [3-540-55296-0]

Vol. 178: Zolésio, J.P. (Ed.)
Boundary Control and Boundary Variation. Proceedings of IFIP WG 7.2 Conference, Sophia-Antipolis, France, October 15-17, 1990
392 pp. 1992 [3-540-55351-7]

Vol. 179: Jiang, Z.H.; Schaufelberger, W.
Block Pulse Functions and Their Applications in Control Systems
237 pp. 1992 [3-540-55369-X]

Vol. 180: Kall, P. (Ed.)
System Modelling and Optimization. Proceedings of the 15th IFIP Conference, Zurich, Switzerland, September 2-6, 1991
969 pp. 1992 [3-540-55577-3]

Vol. 181: Drane, C.R.
Positioning Systems - A Unified Approach
168 pp. 1992 [3-540-55850-0]

Vol. 182: Hagenauer, J. (Ed.)
Advanced Methods for Satellite and Deep Space Communications. Proceedings of an International Seminar Organized by Deutsche Forschungsanstalt für Luft-und Raumfahrt (DLR), Bonn, Germany, September 1992
196 pp. 1992 [3-540-55851-9]

Vol. 183: Hosoe, S. (Ed.)
Robust Control. Proceesings of a Workshop held in Tokyo, Japan, June 23-24, 1991
225 pp. 1992 [3-540-55961-2]

Vol. 184: Duncan, T.E.; Pasik-Duncan, B. (Eds)
Stochastic Theory and Adaptive Control. Proceedings of a Workshop held in Lawrence, Kansas, September 26-28, 1991
500 pp. 1992 [3-540-55962-0]

Vol. 185: Curtain, R.F. (Ed.); Bensoussan, A.; Lions, J.L.(Honorary Eds)
Analysis and Optimization of Systems: State and Frequency Domain Approaches for Infinite-Dimensional Systems. Proceedings of the 10th International Conference, Sophia-Antipolis, France, June 9-12, 1992.
648 pp. 1993 [3-540-56155-2]

Vol. 186: Sreenath, N.
Systems Representation of Global Climate Change Models. Foundation for a Systems Science Approach.
288 pp. 1993 [3-540-19824-5]

Vol. 187: Morecki, A.; Bianchi, G.; Jaworeck, K. (Eds)
RoManSy 9: Proceedings of the Ninth CISM-IFToMM Symposium on Theory and Practice of Robots and Manipulators.
476 pp. 1993 [3-540-19834-2]

Vol. 188: Naidu, D. Subbaram
Aeroassisted Orbital Transfer: Guidance and Control Strategies
192 pp. 1993 [3-540-19819-9]

Vol. 189: Ilchmann, A.
Non-Identifier-Based High-Gain Adaptive Control
220 pp. 1993 [3-540-19845-8]

Vol. 190: Chatila, R.; Hirzinger, G. (Eds)
Experimental Robotics II: The 2nd International Symposium, Toulouse, France, June 25-27 1991
580 pp. 1993 [3-540-19851-2]

Vol. 191: Blondel, V.
Simultaneous Stabilization of Linear Systems
212 pp. 1993 [3-540-19862-8]

Vol. 192: Smith, R.S.; Dahleh, M. (Eds)
The Modeling of Uncertainty in Control Systems
412 pp. 1993 [3-540-19870-9]

Vol. 193: Zinober, A.S.I. (Ed.)
Variable Structure and Lyapunov Control
428 pp. 1993 [3-540-19869-5]

Vol. 194: Cao, Xi-Ren
Realization Probabilities: The Dynamics of Queuing Systems
336 pp. 1993 [3-540-19872-5]

Vol. 195: Liu, D.; Michel, A.N.
Dynamical Systems with Saturation Nonlinearities: Analysis and Design
212 pp. 1994 [3-540-19888-1]

Vol. 196: Battilotti, S.
Noninteracting Control with Stability for Nonlinear Systems
196 pp. 1994 [3-540-19891-1]

Vol. 197: Henry, J.; Yvon, J.P. (Eds)
System Modelling and Optimization
975 pp approx. 1994 [3-540-19893-8]

Vol. 198: Winter, H.; Nüßer, H.-G. (Eds)
Advanced Technologies for Air Traffic Flow Management
225 pp approx. 1994 [3-540-19895-4]

Vol. 199: Cohen, G.; Quadrat, J.-P. (Eds)
11th International Conference on Analysis and Optimization of Systems – Discrete Event Systems: Sophia-Antipolis, June 15–16–17, 1994
648 pp. 1994 [3-540-19896-2]

Vol. 200: Yoshikawa, T.; Miyazaki, F. (Eds)
Experimental Robotics III: The 3rd International Symposium, Kyoto, Japan, October 28-30, 1993
624 pp. 1994 [3-540-19905-5]

Vol. 201: Kogan, J.
Robust Stability and Convexity
192 pp. 1994 [3-540-19919-5]

Vol. 202: Francis, B.A.; Tannenbaum, A.R. (Eds)
Feedback Control, Nonlinear Systems, and Complexity
288 pp. 1995 [3-540-19943-8]

Vol. 203: Popkov, Y.S.
Macrosystems Theory and its Applications: Equilibrium Models
344 pp. 1995 [3-540-19955-1]

Vol. 204: Takahashi, S.; Takahara, Y.
Logical Approach to Systems Theory
192 pp. 1995 [3-540-19956-X]